《光幻湖山——颐和园夜景灯光艺术鉴赏》编委会

主编单位　北京市颐和园管理处
承编单位　《中国建筑文化遗产》编辑部

主　　　任	刘耀忠　李国定
委　　　员	丛一蓬　杨宝利　杨　静　秦　雷　周子牛　王　馨　吕高强
主　　　编	丛一蓬　秦　雷
策　　　划	丛一蓬　秦　雷　金　磊　李　沉
副　主　编	荣　华　李　沉
执 行 主 编	荣　华
参 编 人 员	陈　曲　王　晨　安　静　朱　颐　张　斌　王　娟
摄 影 编 辑	张晓莲　王　晨　张　斌

执行编辑部

文 字 编 辑	金　磊　崔　勇　李　沉　董晨曦　苗　淼　殷力欣　朱有恒　季也清
建 筑 摄 影	中国建筑学会建筑摄影专业委员会
	李　沉　陈　鹤　柳　笛　叶金中　朱有恒　冯新力　毕　楠　等
	北京市颐和园管理处
	张晓莲　周　利　杨晓峰　关　宇
美 术 编 辑	董晨曦　朱有恒　胡珊瑚
翻 译 机 构	中译语通科技股份有限公司

序

/ 单霁翔

从《中国建筑文化遗产》编辑部获悉，他们正在为世界文化遗产颐和园古典园林亮化技术承编《光幻湖山——颐和园夜景灯光艺术鉴赏》一书，我翻阅了书的纲目，感到内容新颖、图文并茂，是一部从新视角反映世界文化遗产地保护与"活态"利用普惠公众的佳作。

本人自 2002 年到国家文物局工作后，就十分关注全国的世界遗产保护工作，尤其对颐和园这座中国古典皇家园林的典范更为重视。早在 2004 年，国家文物局就支持天津大学建筑学院与北京市颐和园管理处所承担的北京科委社会发展项目"颐和园古典园林夜景照明工程技术研究与示范"，2008 年国家文物局也提出了颐和园夜景照明的保护范围的要求，从而确定了保护与"活态"利用的原则与方针。2005 年，我在《加强世界文化遗产保护管理工作的思考》一文中重申了应如何落实"世遗"的保护原则，如对于"保护为主原则"，应明确不以牺牲和破坏世界文化遗产为代价进行开发利用，以换取暂时利益；对"集体保护原则"，应建立一个具有现代科学方法的永久性制度，通过宣传教育增强公众对遗产的自觉尊重；对"真实性原则"，应最大限度地保存所蕴含的全部历史信息，保护和展示"世遗"的本来历史风貌；对"完整性原则"，应在指定文物本体保护计划和范围的基础上，充分重视"世遗"各组成部分之间内在的有机联系等。

《光幻湖山——颐和园夜景灯光艺术鉴赏》一书是很有特点的。其一，它从古典皇家园林文化视角出发，给出了光文化历史底蕴背景下的审美分析，不仅有文化支撑，还有翔实的科学研究作为基础；其二，该书立足于颐和园夜景灯光的艺术赏析，因而在展示必要照明技术的同时，充分利用文学语言，用摄影的手段，表现夜景下的颐和园之美，给读者一个如梦幻般全新的感受；其三，颐和园作为"世遗"地，如何保护，如何有效利用，如何既照亮"世遗"体现皇家园林的博大精深又不造成对古建园林的破坏，在书中都有细致的分析;其四,尤为可贵的是,本书还提出了遵循《世界遗产公约》条款下的《世界遗产地保护与利用照明设计管理条例》,期望在总结颐和园照明工程的基础上,对全国"世遗"保护单位有些借鉴意义。

有鉴于此,我祝贺该书的出版,并希望《光幻湖山——颐和园夜景灯光艺术鉴赏》一书所描绘的颐和园夜景照明技术及传播方式可以推广,让中国更多的世界遗产及文保单位"亮起来""美起来",造福社会与人民。是为序。

中国文物学会会长
故宫博物院院长
2017年10月

1
光耀颐和

Foreword

Shan Jixiang

Upon learning from the China Architectural Heritage editorial board that they were working — as commissioned — on *Glittering Lakes and Hills: An Appreciation of Landscape Lighting at Summer Palace*, a book on illumination techniques applied in the classical gardens of the World Heritage Site of Summer Palace, I browsed the outline of the book and found that the book was novel in content, with an abundance of illustrations. It is a good piece that deals from a brand new angle with the preservation of a World Heritage Site and its "live" utilization for the good of the public.

Since my appointment in 2002 to the State Administration of Cultural Heritage (SACH), I have paid great attention to the preservation of the World Heritage Sites across the country, especially Summer Palace — the exemplar of classical imperial gardens in China. The SACH endorsed way back in 2004 the Beijing Municipal Science & Technology Commission program, "Landscape Lighting Research and Demonstration for Classical Gardens of Summer Palace", undertaken by the Tianjin University's School of Architecture and Beijing Summer Palace Management Office; in 2008 the SACH proceeded to make requirements on the protective scope of Summer Palace landscape lighting and set forth the principles and guidelines of preservation and "live" utilization. In my 2005 article, "Thoughts on Strengthening the Preservation and Management of World Cultural Heritage", I talked once more about how to implement the principles of the preservation of World Heritage Sites. For instance, the protection-first principle should make it clear not to seek development and utilization for short-term interests at the price of destructing World Cultural Heritage; the collective-protection principle requires having in place a permanent system of modern scientific methods to raise public awareness of cultural heritage through education and communication; the authenticity principle requires maximally preserving the entirety of historical information contained in a World Heritage Site, protecting and exhibiting its true historical looks; and the integrity principle requires giving full respect to inherent correlations between the components of a World Heritage Site while preserving its entirety as demarcated.

The book, *Glittering Lakes and Hills: An Appreciation of Landscape Lighting at Summer Palace*, has several impressive features. First, it takes the culture of classical imperial gardens as its point of departure and provides an aesthetic analysis that is not only steeped in history of garden lighting culture but grounded in meticulous studies. Second, as primarily an artistic appreciation of landscape lighting at Summer Palace, the book, while displaying illustration techniques where necessary, makes full use of the literary language and photographic techniques to show the readers a dreamlike brand new experience with the imposing beauty of Summer Palace illuminated at night. Third, the book gives a thorough analysis on how to make the best of Summer Palace as a World Heritage Site and how to illuminate it in a way that displays the great profundity of an imperial garden without causing damage to it. And fourth, it is particularly praiseworthy that the book also proposes "Rules of Lighting Design Management in World Heritage Preservation and Utilization" or "Guidelines", which is consistent with the Convention Concerning the Protection of the World's Cultural and Natural Heritage, with a view to shedding light on world heritage preservation efforts across China based on experience learned from Summer Palace lighting program.

Hence, I extend congratulations on the publication of the book. And I hope that Summer Palace lighting techniques described therein may be applied elsewhere in China, thus illuminating and beautifying more cultural heritage sites in the country and benefiting the public. The above is my foreword for the book.

Shan Jixiang
President of the Chinese Society of Cultural Heritage
Director of the Palace Museum
October 2017

前言

刘耀忠

颐和园是中国乃至全人类的文化珍宝，作为1998年进入联合国教科文组织《世界遗产名录》的世界文化遗产，近20年来它遵循《世界遗产公约》的要求，在古典皇家园林保护诸方面作出不懈努力和探索。其中，颐和园古典园林夜景照明既是传承又是"活态"利用与发展的创新。

颐和园（前身为清漪园），始建于公元1750年，时值中国最后一个封建盛世——"康乾盛世"。它集传统造园艺术之大成，既包含中国皇家园林的富丽恢弘，又处处显现"虽由人作，宛自天开"的意境。如何让颐和园在夜色中继续传承华夏古建园林的魅力，如何在构建起美好绚丽的光环境后不对"世遗"产生破坏作用，如何将颐和园光文化的历史底蕴呈现给公众并为北京增添新景……都成为北京市颐和园管理处面临的新课题。为贯彻《世界遗产公约》及《中华人民共和国文物保护法》，为了体现科学性、生态性、安全性的文物保护照明原则，早在2004年北京市科委立项资助成立了由天津大学建筑学院、北京市颐和园管理处联合组成的"颐和园古典园林夜景照明技术研究及示范"课题组，历时五年，全面依据科学实验，在多方面取得了具有颐和园古典园林夜景照明特色的技术参数与设计建设要点。

《光幻湖山——颐和园夜景灯光艺术鉴赏》一书，是在2009年颐和园照明研究及2016年颐和园古典夜景照明评估等项目基础上完成的。在北京市颐和园管理处各级管理部门指导及配合下，中国建筑学会建筑摄影专业委员会、《中国建筑文化遗产》编辑部同人组织文保专家、建筑摄影、文字编辑、版式设计等创作团队，经过不懈努力使颐和园照明技术研究的成果以一种普及"读本"的形式面向读者。需要说明的是，在颐和园古建园林艺术欣赏的数十本图书中，图文并茂，将摄影视觉与文学笔触相结合解读鉴赏的图书这还是第一册。尽管我们与承编方为此作出了不懈努力，但难免还有不妥之处，还有待在未来修订时，再创作、再提升。

《光幻湖山——颐和园夜景灯光艺术鉴赏》一书除形式新颖,编排轻松,图文精湛外,颐和园管理处还始终要求对"世遗"项目的"亮化"手段要千方百计突出"保护为先"的原则,科学合理"活态"利用且使照明手段技术可逆;此外主编、承编单位还共同在项目编撰过程中大胆探索,并提出了对全国52处"世遗地"有适用价值的"点亮"世界遗产"技术导则",以求为中国世界遗产的保护与管理技术作出贡献。

值此《光幻湖山——颐和园夜景灯光艺术鉴赏》出版之际,感谢中国文物学会会长、故宫博物院院长单霁翔长期以来对颐和园保护的指导帮助以及对本书所作的序,也感谢为本书编辑出版作出贡献的所有专家与编辑同人,特表敬意。

颐和园园长
2017年10月

1 石舫夜色

Preface

Liu Yaozhong

Summer Palace, or Yiheyuan, is a cultural gem of China and even of the world as a whole. Over nearly 20 years since its enrollment in 1998 as a UNESCO World Heritage Site, unremitting efforts have been made, as required by the Convention Concerning the Protection of the World's Cultural and Natural Heritage, to protect it as a classical imperial garden. As part of those efforts, landscape lighting for the classical gardens within Summer Palace is intended not only as a means of inheriting but as an innovation in "live" utilization and development.

Construction of Summer Palace, initially called "Qingyi Garden" (Garden of Clear Ripples), began in 1750, in the last heyday of Chinese feudalism — "Kangqian Era of Prosperity". It is a great paradigm of traditional gardening, majestic and magnificent as an imperial garden, with its every part appearing as if "created by nature". How to let Summer Palace continue its charm at night as an age-old garden, how to avoid damage to the World Heritage Site after creating a beautiful light environment for it, how to convey to the public the historical significance of its lighting culture and add splendor to Beijing... all these are new challenges the Beijing Summer Palace Management Office is facing. To implement the Convention Concerning the Protection of the World's Cultural and Natural Heritage and the Law of China on the Protection of Cultural Heritage and observe the principles of scientificity, ecology, and safety with regard to the use of lighting at a cultural heritage site, back in 2004 the Beijing Municipal Science & Technology Commission funded the founding of the task force for Landscape Lighting Research and Demonstration for Classical Gardens of Summer Palace, which consisted of the Tianjin University's School of Architecture and the Beijing Summer Palace Management Office. In the five years that followed, this task force, based on scientific experiments, developed the technical parameters and the main points of design and construction with the characteristic of landscape lighting for the classical gardens of Summer Palace.

The book, "Glittering Lakes and Hills: An Appreciation of Landscape Lighting at Summer Palace", was written on the basis of the 2009 Summer Palace lighting research program and the 2016 Summer Palace garden lighting evaluation. Intended as a popular "primer" on the outcomes of the technical research into lighting for Summer Palace, it was completed through concerted efforts of the professional teams of cultural heritage specialists, architectural photographers, editors and format designers, as organized by the Architectural Photography Committee of the Architectural Society of China and the China Architectural Heritage Editorial Board with guidance and collaboration from various divisions of the Beijing Summer Palace Management Office. It is worth noting that, of the books on classical gardens of Summer Palace — which number several tens, this book is the first of a kind that is rich in literature and photography. Despite our editorial partner's meticulous efforts, the book may have deficiencies that need to be corrected or improved in the future.

Besides producing this book that is refreshingly novel in form and format and fine in content, the Beijing Summer Palace Management Office always insisted that the "illuminating" means for a World Heritage Site should give top priority to preservation and be scientifically suitable and technically reversible. Moreover, our editorial partners also made bold explorations in the process of compilation and proposed "technical guidelines" of a universal value for the illumination of 52 World Heritage Sites across China, with a view to contributing to the preservation and management of World Heritage Sites in the country.

We'd like to take this opportunity to express special thanks to Mr. Shan Jixiang, President of the Chinese Society of Cultural Heritage and Director of the Palace Museum, for his persistent guidance and assistance to Summer Palace and for the foreword he wrote for this book, as well as to all experts and editors who contributed to the compilation and publication of the book.

Liu Yaozhong
Director of Beijing Summer Palace Management Office
October 2017

篇·三

164　颐和园夜景照明分析与启示

166　● 颐和园皇家园林保护工程

170　● 颐和园夜景照明设计框架

174　● 照明设计技术要点

178　● 世界遗产地保护性照明设计施工建设管理条例（建议稿）

185　拍摄与编后记

目录

篇·一

- 4 序·单霁翔
- 8 前言·刘耀忠
- 24 颐和园的夜间照明钩沉

篇·二

- 35 颐和园夜景照明欣赏
- 36 概述
- 42 谐趣园
- 56 霁清轩
- 70 苏州街
- 84 后溪河
- 92 长廊
- 106 万寿山·佛香阁
- 122 西堤
- 136 九道弯·知春亭·文昌阁
- 150 十七孔桥·南湖岛·廊如亭

Table of Contents

Foreword   Shan Jixiang / 4

Preface   Liu Yaozhong / 8

Chapter One   Historical Anecdotes about Summer Palace Illumination at Night / 24

Chapter Two   Illuminated Nightscape of Summer Palace / 35

    Overview / 36
    Garden of Harmonious Pleasures / 42
    Jiqing Pavilion / 56
    Suzhou Street / 70
    Houxi River / 84
    Long Corridor / 92
    Longevity Hill and Tower of Buddhist Incense / 106
    West Dike / 122
    Nine Bay, Zhichun Pavilion, and Wenchang Tower / 136
    Seventeen-Arch Bridge, Nanhu Island, and Kuoru Pavilion / 150

Chapter Three   Summer Palace Landscape Lighting Analysis and Enlightenment / 164

    Program for the Preservation of Summer Palace Imperial Gardens / 166
    Design Framework for Summer Palace Landscape Lighting / 170
    Technical Principles of Lighting Design / 174
    Management Rules for Protective Lighting Design and Construction at World Heritage Sites (Proposal Manuscript) / 178

Photography and Postscript / 185

篇·一

颐和园夜间照明钩沉
Historical Anecdotes about Summer Palace Illumination at Night

颐和园的皇家园林之美举世无双,其夜间照明更是反映了时代的发展。从幽暗、微弱的火烛、油灯,到五彩斑斓的夜景灯光,从以往的陶、竹、铜、玉,到今日的玻璃、合金、卤钨、LED,颐和园灯具的历史变迁为"光幻湖山"增添了新的注解。无论白昼还是夜晚,似歌如画的美景韵味无穷,令人难以忘怀。让我们从历史走来,感受绝妙的照明艺术。

The beauty of the imperial gardens in Summer Palace is unique in the world, and its illustrated landscape reflects the progress of times. From the dim candle lights through the oil lamp to the colorful lights, and from the pottery, bamboo, copper and jade to glass, alloy, halogen tungsten and LED, the evolution of the lighting in Summer Palace has added new annotations to the Glittering Lakes and Hills. Be it day or night, Summer Palace is always hauntingly beautiful. Let us go to feel wonderful lighting art from the historical anecdotes about Summer Palace illumination at night.

有关颐和园园林建筑艺术的佳话家喻户晓，但涉及颐和园古代夜间照明的问题则鲜见，即便学术界及硕士、博士学位论文有涉及颐和园夜间照明的研究论题，也都是关于颐和园的现代照明技术研讨。在此钩沉一番颐和园夜间照明脉络（止于1904年），旨在秉承传统并发扬光大、推陈出新。

据史料记载：自1879年爱迪生发明电灯以后，电光源逐渐在世界各地普及，也及时传布到中国。光绪十六年（1890年），在颐和园的东宫门外右侧建有一小型发电厂（即颐和园电灯公所，该所与城内西苑电灯公所同为北京最早的发电供电设施）专门供给颐和园夜间电灯照明。始料不及的是光绪二十六年（1900年）八国联军入侵北京而导致西苑、颐和园电灯公所两套发电机组及电灯设备全部遭到破坏，直至光绪三十年（1904年），西苑、颐和园电灯公所才得以恢复发电供电，电灯才在西苑、颐和园重新照亮。此前的园中照明均由宫灯中燎、炬、烛等烛光与油灯提供。

生活于农业文明国度的中国古人遵循"日出而作，日落而息"的生活规律，"明而动，晦而休，无日以怠"则成为人与自然和谐共处并顺应自然的生态文明智慧的显现。故自刀耕火种的远古至隋唐漫长的历史时空中，中国古人的夜间活动相对于欧洲人激情满怀的夜生活显得沉寂许多，甚至说夜幕中除传宗接代外，则是贫乏阙如的。唐代以降，始出现上元节（农历正月十五）、中元节（农历七月十五）、下元节（农历十月十五）三个夜间街市，其余的夜间作息均实行宵禁。直到宋元时期，由于城市经济与文化生活发展而取消里坊制，加之勾栏瓦舍、舞榭歌台及丰富的市井文化的出现导致夜间生活变化之后，夜间照明在中国城市才逐渐兴盛起来。辛弃疾词《青玉案·元夕》便是写照——"东风夜放花千树。更吹落，星如雨。宝马雕车香满路。凤箫声动，玉壶光转，一夜鱼龙舞。蛾儿雪柳黄金缕，笑语盈盈暗香去。众里寻他千百度，蓦然回首，那人却在灯火阑珊处。"但一直以来中国古代夜间通常通过燃烧火烛、油灯来驱除黑暗而带来光明，并且仅在特别的佳节或喜庆的时刻才会有千姿百态的灯火光彩照明。这种夜间情状即便是在作为帝都的北京也是如此。

农历正月十五是民间传统的上元节，也叫元宵节、灯节，所谓"正月十五闹花灯"。整个过程

1
仁寿殿里的电灯

2
慈禧居所乐寿堂内的电灯

从正月十三持续至正月十七，其中正月十三为上灯、正月十四为试灯、正月十五为正灯、正月十七为罢灯。于是乎张灯、赛灯、观灯也就逐渐成为上自宫廷下至民间热闹非凡的传统习俗。每年的七月十五是中元节，它是与正月十五的上元节和十月十五的下元节鼎足而立的传统节日，在道教中有"三元""三官"的别称。上元节又称"上元天官节"，系上元赐福天官紫微大帝诞生；中元节又称"中元地官节"，系中元赦罪地官清虚大帝诞生；下元节又称"下元水官节"，系下元解厄水官洞阴大帝诞生。道教《太上三官经》云："天官赐福，地官赦罪，水官解厄，一切众生皆是天、地、水三官统摄。"上元节、中元节、下元节无疑成了灯节。上元节、中元节、下元节时，颐和园张灯结彩，夜间照明是必不可少的仪礼活动。从颐和园亭台楼阁桥船舫街坊以及山石路径等沿用的各类灯饰来看，中国古代传承的由陶、青铜、铜、铁、银、鎏金、锡、玉、木、竹、石、玻璃、合金、瓷等材料制作而成的各种类型灯饰均——可见，由此不难料想颐和园夜间华灯之下丰富多彩、摇曳多姿的韵致。

1
文昌阁旁建了颐和园首个发电机（图中烟囱位置）

2 3
万寿山的古今对比。晚清时，水木自亲殿前的探海灯杆高悬着一盏电灯，灯杆上绘金色云龙，上托半圆形透雕龙纹的镀金铜梁

颐和园分为宫廷生活区、万寿山风景游览区、昆明湖景区三大部分。宫廷生活区包括以仁寿殿为中心的朝政建筑以及玉澜堂、乐寿堂等。万寿山风景游览区包括万寿山前山宗教建筑、万寿山西部景区、万寿山后山景区、万寿山东部景区。昆明湖景区包括昆明湖东堤景区、昆明湖西堤景区、昆明湖西岸景区、耕织图景区。美学家叶朗在《中国美学史纲》中记载，明清园林美学的中心内容，是园林意境的创造和欣赏。中国古典美学的"意境说"在园林艺术、园林美学中得到了独特的体现。在一定意义上可以说，"意境"的内涵，在园林艺术中的显现，比较在其他艺术门类中的显现，要更为清晰，从而也更容易把握。园林建筑是人所栖息的空间艺术，颐和园景区白天的园林建筑景象由亭台楼阁、湖光山色组成，而夜间则只有通过自然的天光云影以及油灯、烛光的光亮映照园林建筑区域才得以形成。颐和园园林建筑照明涉及各类不同的建筑照明、水体照明、花卉照明、树木照明、山石照明、道路照明以及室内居住照明等方面，包括如下几种典型的夜间照明模式，即庄重朴实模式（如房屋建筑与叠山理水及后山买卖街）、私密宁静模式（通幽曲径与月色天光及闺密私房）、华丽宏大模式（如亭台楼阁与桥舫及戏台）、神圣庄严模式（如宫殿与庙堂）、古朴

诗意模式（四季自然风光与远山景观）、梦幻浪漫模式（如昼夜日月星光）等。这些照明模式的灯光以其强弱、疏密、浓淡、错落等光彩与声、色、星光构成园林之美景。

唐代白居易《琵琶行》记有："浔阳江头夜送客，枫叶荻花秋瑟瑟。……移船相近邀相见，添酒回灯重开宴，千呼万唤始出来，犹抱琵琶半遮面，转轴拨弦三两声，未成曲调先有情。……别有幽愁暗恨生，此时无声胜有声。"杜牧《泊秦淮》写有："烟笼寒水月笼沙，夜泊秦淮近酒家；商女不知亡国恨，隔江犹唱《后庭花》。"张继《枫桥夜泊》记有："月落乌啼霜满天，江枫渔火对愁眠；姑苏城外寒山寺，夜半钟声到客船。"李商隐《夜雨寄北》写有："君问归期未有期，巴山夜雨涨秋池；何当共剪西窗烛，却话巴山夜雨时。"更有宋代西湖老人《西湖老人繁胜录·街市点灯》有言："庆元间，油钱每斤不过一百会，巷陌爪扎，欢门挂灯，南至龙山、北至北新桥，四十里灯光不绝。城内外有百万人家，前街后巷，僻巷亦然。"明代李渔《闲情偶寄·居室部》将泛舟荡漾于园林建筑与湖光山色中的情景描绘成画境文心之境——"坐于其中，则两岸之湖光山色、寺观浮屠、云烟竹树，以及往来樵人牧竖、醉翁游女，连人带马，尽入便面之中，作我天然图画。且又时时变幻，不为一定之形。非特舟行之际，摇一橹，变一象，撑一篙，换一景，即系缆时，风摇水动，亦刻刻异形。是一日之内，现出百千万幅佳山佳水，总以便面收之。……此船窗不但娱己，兼可娱人。不特以舟外无穷之景色摄入舟中，兼可以舟中所有之人物，并一切几席杯盘射出窗外，以备来往游人之玩赏。何也？以内视外，固是一幅便面山水，而以外视内，亦是一幅扇头人物，……无一不同绘事。同一物也，同一事也，……人人俱作画图观矣。"照理来说作为堂堂的皇家苑囿的颐和园逢年过节、喜庆活动以及歌舞台榭之际必然有张灯结彩的夜间照明景致，遗憾的是颐和园鲜有类似朱自清描绘六朝古都金陵秦

1
罗布林卡壁画中的颐和园

2
清代《三山五园图》

3
晚清颐和园全图

淮河畔灯光水影的《桨声灯影里的秦淮河》那样的美文来记载幽燕帝都西郊颐和园的水色山光灯影盛况以及华灯映水、画舫凌波、树影婆娑、亭台楼阁影影绰绰的朦胧光景。清人郭嵩焘撰有《清漪园记》，也只是流水账式地记录颐和园的园林建筑、昆明湖、万寿山等处景观布局而无涉颐和园的夜色照明。或许是作为帝王将相之园，其私密处当密封而不可泄露，有关颐和园的夜间照明的史料难得一见。清代皇帝弘历对颐和园情有独钟并写有一千多首有关颐和园的诗词，笔触多涉猎颐和园及借景西山与香山白天的湖光山色、四季变换以及亭台楼阁错落有致的分布，少有描绘夜间照明下的颐和园迷人的景致。学术界对有关颐和园古代夜间照明问题则显得语焉不详，这无疑给后人研究颐和园夜间照明带来困惑。颐和园以自然山水景致为主体、以人工造景为辅综合而成的皇家园林所显示的恢弘磅礴的气势和山明水秀的融建筑的形态、情态及生态为一体的文化生态美是显而易见的，而夜幕下依靠传统的燎、炬、烛、灯照明之下的颐和园到底是怎样的景致还是一个谜团。

The garden architecture of Summer Palace is well known, but the landscape lighting in the ancient Summer Palace is rarely discussed, and the academics, the doctor and master candidates mostly study the modern lighting technology of Summer Palace if their topics are relevant to landscape lighting. To fill the research gap, this article presents the evolution of the lighting of the ancient Summer Palace till 1904.

According to historical records, since Edison invented the electric light in 1879, electric light sources have gradually spread to all parts of the world including China. In the 16th year of Guangxu's reign (1890), outside the East Palace of Summer Palace was a small power plant, i.e., Summer Palace Lamp House, which together with the Xiyuan Electric Lamp House, represented the earliest power supply facilities in Beijing, in the service of the nighttime lighting of Summer Palace. Unfortunately, the Eight-Power Allied Forces invaded Beijing in the 26th year of Guangxu's reign (1900), leading to the destruction of the two generator sets and all lighting equipment in the Xiyuan and Summer Palace lamp houses. It was not until the 30th year of Guangxu's reign (1904), the two lamp houses were restored and the lights of Xiyuan and Summer Palace were thus relit. The electric lamp in the Hall of Benevolence and Longevity of Summer Palace was installed in 1904, powered by a coal-fired generator from Summer Palace Lamp House. Historically, it should be the first ever electric lamp in Beijing. Before that Summer Palace's lighting depended on fire, torches, candles and oil lamps.

The ancient Chinese people in the agricultural country followed the routine of "rising with the sun and going to sleep when it gets dark"; and the record "to work while the day is bright,

1
1864年的玉泉山

2
如今的玉泉山

to sleep when the day dims, and never to idle away a single day" reflects the harmonious coexistence of man and nature and the human wisdom of respecting nature. Therefore, for so long a time from the remote antiquity dominated by the slash-and-burn cultivation to the Sui and Tang dynasties, the night life of Chinese people was quiet compared with the nocturnal activities of their European contemporaries. Or rather, the people in ancient China led a quite monotonous night life. Starting from the Tang Dynasty, the night curfew was implemented except for the Lantern Festival (the fifteenth day of the first lunar month), the Ghost Festival (the fifteenth day of the seventh lunar month) and the Spirit Festival (the fifteenth day of the tenth lunar month). It was not until the Song and Yuan dynasties that the neighborhood system was abolished along with the economic and cultural development of cities, which coupled with the emergence of entertainment centers and set-ups and increasingly colorful city life gave rise to changes in nightlife. As a result, nighttime lighting gradually became popular in cities. Xin Qiji's poem "The Lantern Festival" depicts such a scene, "During the night the east wind blew open thousands of sliver flowers and blew down fireworks like stars and raindrops on the floor. Painted carriages and precious horses bustled back and forth while fragrance filled the road. Sweet music was played high, and a bright moon hung in the sky. Fish-like and dragon-like lanterns danced merrily the whole night. Pretty women wore ornaments of various kinds on their heads. They chatted cheerfully and laughed lightheartedly, leaving secret fragrance behind. In the crowd for a thousand times, I failed to find my love. Suddenly I turned back and saw her in the corner where lights were sparse and somber". But the people in ancient China usually used candles and oil lamps to dispel darkness and would put on myriads of lights on festive days only. That was also the case with the capital city Beijing.

The fifteenth day of the first lunar month is the traditional Shangyuan Festival, also known as the Lantern Festival, as reflected in the saying "lantern show on the fifteenth day of the first lunar month". Regarding the period from January 13 to January 17 of the lunar calendar, the

13th day was for the start of the lantern show, 14th trial operation, 15th formal operation and 17th end of the lantern show. Therefore, it has become a hilarious tradition to decorate, show and admire the lights from the imperial palaces to common streets. The Ghost Festival which falls on July 15 in the lunar calendar is one of three parallel festivals, with the other two being the Lantern Festival (January 15 of the lunar calendar) and the Spirit Festival (October 15 of the lunar calendar). The Ghost Festival is also a Taoist festival , called the three-yuan and three-guan in Taoism. The Lantern Festival is also known as "Shangyuan Heavenly God Festival" when God Ziwei, the Heavenly God who blesses the people , was born; the Ghost Festival is also known as "Zhongyuan Earth God Festival" when God Qingxu ,the Earth God who pardons the people , was born; the Spirit Festival is also known as "Xiayuan Water God Festival" when God Dongyin ,the Water God dispels disasters , was born. As the Doctrine of Taoist Three Deities goes, "Heavenly God blesses the people, Earth God pardons the people and Water God dispels disasters. All beings are under the control of the three gods". The Lantern Festival, the Ghost Festival and the Spirit Festival have undoubtedly become festivals of lanterns. During these festivals, it is an indispensable part of the ritual activities in Summer Palace to decorate and hang the lights. The lamps and lanterns passed down from old days in Summer Palace, as seen in the pavilions, belvederes, boats, bridges, streets and paths come in varied patterns and are made of diverse materials like pottery, bronze, copper, iron, silver, gold, copper, iron, silver, gilding, tin, jade, wood, bamboo, stone, glass, alloy, and porcelain. So, it can be imagined what a marvelous night landscape unfolded itself with so many lamps in diverse forms and materials lit.

Summer Palace is divided into three parts: the imperial living area, Longevity Hill tour zone and Kunming Lake scenic area. The imperial living area includes the royal court buildings centered on Qinzheng Palace, and Yichun Hall, Yulan Hall and Leshou Hall. The Longevity Hill tour zone includes the religious buildings in front of the Longevity Hill, the western scenic spot, the rear scenic spot and the eastern scenic spot of the Longevity Hill. The Kunming Lake scenic area includes the East Dyke Scenic Spot, the West Dike Scenic Spot, the Western Bank Scenic Area, and Farming Area. In the "Outline of Chinese Aesthetics", the esthetician Ye Lang observed, It is central to the garden aesthetics in the Ming and Qing dynasties to highlight the creation and appreciation of the artistic conception of garden. The theory on artistic conception of Chinese classic aesthetics is distinctively reflected in the garden art and garden aesthetics. In a sense, the implications of the artistic conception are best manifested in the art of gardening and easier to grasp. The garden architecture is the art about the space inhabited by people. The daytime scenery of Summer Palace is formed by the pavilions, belvederes, lakes and hills whereas the nighttime scenery is formed by the night views of these natural and man-made sights under the radiance of oil lamps and candles. Summer Palace's lighting involves architectural lighting, water lighting, flower lighting, tree lighting, hill lighting, road lighting and indoor lighting. The nighttime lighting styles have been ingeniously designed as follows: the somber and simple style for the buildings, mountains and waterways as well as the marketing street in the rear mountain, the private and tranquil style for the winding paths and boudoirs under moonlight, the magnificent style for pavilions, belvederes, bridges, boats and stages, the sacred and dignified style for the palace and the imperial court, the traditional and poetic style for seasonal natural landscape and remote mountain views, the dreamlike and romantic style for the landscape under the sun, the moon and stars. In sparse or dense concentration, the originally arranged lights of different intensity and brilliance together with the starlight, sounds and colors constitute the beautiful night landscape of Summer Palace.

In the Tang dynasty Bai Juyi's "Pipa Xing" went as follows, "I saw off a friend by the Xunyang River at night. Maple leaves and flowering silver grass rustled in the autumn wind…Manipulating our boat a little closer, we invited her to come over. We lighted up lamps, added more wine, and resumed our dinner. Upon our repeated invitations, she finally agreed to come out and play. Holding the pipa in her arms, she used it to hide half of her face. She first tuned up and plucked the strings two or three times. Before she started, she already revealed what was on her mind… Some hidden sorrows maybe found a way to come around. At a time like this, silence was more fitting than any given sound. " Du Mu's "Mooring on the Qinhuai River " wrote, "The river was cold and enshrouded in a fog, the river sand enshrouded in moonlight; I came to moor on the Qinhuai River after dark where there were cabarets nearby. Courtesans didn't perceive the sorrow of a perished empire; across the river singing of "A Flourishing Backyard" continued into the night." Zhang Ji's "Mooring by Maple Bridge at Night" wrote, "The crows at moonset cried in the frosty sky; facing dim lights on fishing boats neath maples, I lay sad. Outside the city wall of Gusu, the ringing bells from the Temple of Cold Hill broke the passenger's dream in midnight ." Li Shangyin's "Writting on a Rainy Night to My Wife in the North" wrote, "You asked me when could return, but I don't know. It is raining in western hills and autumn pools have overflown. When can we trim by window side the redundant candlewick and talk about the western hills in rainy night? " In the Song dynasty, the West Lake Elder wrote in "West Lake Elder's Record of the Prosperous Scene Ablaze with Lights", "During the Qingyuan Era, the oil cost no more than 100-plus hui per jin. From the Dragon Mountain in the south to the Beixin Bridge in the north, the way wound for 40 li. Along the way, the streets and lanes were festooned with lights. There were millions of households in and outside the city. Everywhere, including the out-of-the-way lanes, was ablaze with lights." In the Ming dynasty, Li Yu in "Pleasant Diversions" depicted his experience of boating in the garden where the architecture was illuminated by the moonlight and lamps, "Sitting in the boat, I saw the scenery on both sides, including the lakes, hills, temples and stupas, trees, bamboo, woodcutters, shepherd boys, drunken men, ladies taking a stroll, and people riding horses. The natural picture was constantly changing. Even when the boat was moored, a move of the pole or oar and the ripples caused by wind will send the changing scene. Thus, millions of landscapes emerged a day…While we took in the enchanting views outside the boat, everything in the boat like the cups and disks on the table was illustrated and visible from outside, which amused the fellow tourists. Why? The same articles and things were all part of the picture". It is reasonable to imagine that Summer Palace as a world famous royal court must have been well illustrated during festivals and holidays and in particular its pavilions and belvederes must have been festooned with colorful lights, but it is a pity that there is hardly any beautiful literature depicting such scenes. Zhu Ziqing's "Oar Sound and Light Shadow on the Qinhuai River"portrays the enchanting scenes by the river. But such literature is lacking to describe the illustrated landscape of Summer Palace in the western suburb of the capital, like the lights reflected in water, the painted pleasure boat on the lake, the shadows of trees dancing in the breeze, and the pavilions and belvederes seen vaguely. Guo Songtao of the Qing dynasty wrote "Record of Qingyi Garden" where he gave a running account of the scenic spots in Summer Palace, like the garden architecture, the Kunming Lake and the Longevity Hill, with no mention of the illustrated landscape therein. It is possible that things about Summer Palace as a royal palace should be kept in secret, which explains why the data about the nighttime lighting of Summer Palace is hardly seen. The Qing Emperor Hongli was especially fond of Summer Palace and wrote over 1,000 poems about Summer Palace, which mostly depict the landscape of Summer Palace together with the Western Hill and the Fragrant Hill, the seasonal views and the distribution of the pavilions and belvederes, but hardly mention the engaging views of the illustrated landscape of Summer Palace. The academic circle discussed little about the nighttime lighting of Summer Palace, which makes the following generations of people puzzled. Summer Place with the natural landscape in the main supplemented by man-made sights presents magnificent views typical of imperial gardens. Its daytime beauty is evident whereas what on earth Summer Palace looked like under the lights of fire, torches, candles and lamps largely remains mysterious.

篇·二

颐和园夜景照明欣赏
Illuminated Nightscape of Summer Palace

颐和园的夜景令人神往。

夕阳伴着天光西沉，颐和园的灯光成为主角。灯光下，佛香阁更显庄严宏伟，十七孔桥愈发梦幻浪漫，谐趣园充满古朴诗意，西堤尽展安宁静谧。

不同的光影、明暗的变化是天光、自然环境与人工环境的完美结合。光与照明是让人们获得夜间景致欣赏的第一要素，最大限度地优化光线给游人带来的心理、生理上的感受，是创造颐和园美好景观的关键。这里不仅要有模拟自然光变化获得的舒适感，还要为欣赏夜景引入犹如"光色与味觉""亮度与视觉"等特有的知觉效果，从而使古老的建筑园林焕发新生机。

The night scene of Summer Palace is particularly fascinating

When the sun disappears below the horizon and twilight falls, the spectacular lights of Summer Palace become the protagonist on the stage against the night scenes. With the lights on, the Tower of Buddhist Incense is solemn and magnificent; the Seventeen-Arch Bridge, dreamy and romantic; the Garden of Harmonious Pleasures, unadorned and poetic; and the West Dike, peaceful and tranquil.

Different changes of light and shade are the results of perfect combination of twilight, natural environment, and artificial environment. Light and illumination make it possible for people to enjoy the scenery, and maximize visitors' physical and psychological feelings of light, which is the key to creating a beautiful landscape for Summer Palace in the dark. Not only does it have to simulate the comfort of changing natural light, but also to introduce the unique effects of perception, such as light color and taste, brightness and vision, for visitors to enjoy the night scenes, revitalizing the ancient landscape garden.

概述 Overview

颐和园以万寿山、昆明湖为基址，以杭州西湖为蓝本，运用对比、借景等中国传统造园方法，将自然环境与人工建造巧妙地融为一体，既显示出帝王园囿雍容磅礴的气势，又富有江南水乡的雅致清新，达到"虽由人作，宛自天开"的境界，有"皇家园林博物馆"之誉。

The area covering Longevity Hill and Kunming Lake is the footing of Summer Palace, and West Lake in Hangzhou is the blueprint. Traditional Chinese landscaping approaches, such as contrasting and view borrowing, among others, were used to skillfully integrate the natural environment and the artificial constructions into an organic whole. The result manifests not only the royal landscape garden's grace and magnificence, but also its richness of the elegance and briskness of the patchwork of waterways in Jiangnan, reaching the state of "works of nature, shaped by human" and enjoying the reputation of "the Museum of Royal Gardens".

被称作"佛山"的万寿山，山势连绵起伏，东西两坡舒缓而对称；"寿海"昆明湖以环抱之势，广纳天下奇观。暮色四合，晚烟上腾，被雾气裹挟的东堤伴着今日最后的鸽哨，在长空下显露出迷人的曲线。廊桥上璀璨的灯光，似银河的映像，静默地沉寂在昆明湖水中，时而被周身嬉闹的鱼群追逐起波涛，也只是翻一个滚儿，又晃荡成一条朦胧的光带。

The east and west slopes of the rolling Longevity Hill, also known as "Buddhist Hill," are gentle and symmetrical. Kunming Lake, known as "Sea of Longevity," embraces the spectacles of land under heaven. While the twilight glow of the sky is fading and the late mist is rising up, the East Dike, enveloped

1
佛香阁

in a blanket of fog and accompanied by the last song of pigeon whistle one day, reveals its charming curve under the vast sky. Glittering lights on the corridor bridge, like the image of the Milky Way, silently reflect in the lake of Kunming, and now and then, after rolling waves made by groups of frolicking fish chasing each other died away, become a hazy light ribbon sloshing around.

夜景下的颐和园有着别样的魅力。无论是临西山远眺与美丽的京城融为一体的"光耀颐和"，还是移步异景，在昆明湖畔欣赏着斑斓帆影映衬下的佛心莲刹，柳岸花堤间的灯带变幻着无尽的组合，照亮了沉睡百年的佛香阁，用十七孔桥的炫彩连接起南湖岛与东堤，串联起过去和未来。

The night scene of Summer Palace is particularly enchanting. Whether you present yourself on the Western Hill to overlook in the distance the view of "Dazzling Summer Palace" blended nicely with the beautiful capital, or you stroll down various scenes to admire on the lakefront of Kunming Lake the Buddhist temple, you'll find the endless combinations of the ribbon of lights in the midst of willow banks and floral dikes, illuminate the slumbering, century-old Tower of Buddhist Incense, and the Seventeen-Arch Bridge's awesome colors, connect Nanhu Island and the East Dike, linking the past with the future.

晚清时的颐和园并非此番景象。

Summer Palace in the late Qing dynasty did not have this kind of vibrant outlook.

1
霁清轩曲廊

2
霁清轩小品

没有光的夜幕隔绝了追寻光明的眼睛。灯油的光挣扎着想要冲破黢黑的夜空，却只发出如豆的微光。正是夜景灯光的出现，使颐和园更加幽静、迷人，从而在人们面前展现出颐和园特有的意境，在传承中国园林艺术的美学基础上，塑造出具有现代审美意义的新的园林艺术人文景观。

Without lighting in the night, the curtain of night isolated the palace from the pursuing eyes. The light of oil lamps was struggling to break through the darkness under the night sky, but only emitted shimmering, glimmering light. It is precisely the arrival of nightscape lighting that makes Summer Palace more tranquil and charming, thereby showing in front of people the distinctive artistic mood of the palace. On the basis of inheriting the aesthetics of Chinese garden, it has shaped a new humanistic landscape of garden art with modern aesthetic significance.

如想一览颐和园夜景的全貌，就需要按一种起承转合的顺序进行观赏，因为颐和园夜景照明遵循着项目亮化的主辅依次展开，因此本书介绍的观赏线路自然与白日略有不同。当然，这里提供的是参考的游览项"线路"，也仅仅是以点带面。

> If you want to view the panorama of Summer Palace in the night sky, you need to view it in the sequence of introduction, elucidation of the theme, transition to another viewpoint, and summing up, Summer Palace's nightscape lighting follows the descending order of illuminating the principal objects and then the secondary objects. Therefore, the viewing line introduced in this book is slightly different from that of daytime. Of course, what we provide here is a reference "itinerary", which is merely a list of attractions that allow people to figure out how to make a full plan.

需要说明的是点亮颐和园古建园林的夜色，照明艺术功不可没。但更重要的是将光照艺术记录下来的建筑摄影工作，可以算作一个建筑文化传承的"项目"。在自然场地中获取庇护是古今中外建筑师基本的职能，而统领着图像和视觉艺术的建筑摄影师，不仅可将场地转变为"景观"，还可将自然变为"舞台"。颐和园夜景新视角的呈现，是一次完美的重生，它是这座皇家园林世界遗产的光耀，更给市民与中外游客带来前所未有的惊喜。建筑照明夜景摄影超越了建筑本身，也可超越所有风景园林的场域。因为，本书所展现的图片是在既有古建园林风光基础上的再创作，透过一位位摄影师独到的视角，赋予其新的欣赏点。画面的冲击力让建筑凸显出人文内涵以及珍贵遗产价值。颐和园夜景照明的艺术不仅靠硬件支撑，更是软实力的提升，建筑摄影图片已将古建园林、摄影艺术紧紧地融合在一起。下面请跟随我们走进夜色中的颐和园，去感受它那夜幕下特有的迷人意境与风采。

> It should be noted that the contributions of lighting arts cannot go unnoticed to illuminate Summer Palace's nightscape of ancient gardens. But more

importantly, architectural photographers' works of recording light designs can be deemed to be part of cultural inheritance project of architecture. To obtain shelters in the natural field is an architect's job at all times and in all places, and an architectural photographer, who commands the arts of image and visual arts, can not only transform the site into a "landscape," but also transform nature into a "stage." The presentation of new perspective of the nightscape is a perfect rebirth of Summer Palace, which is the glory of this royal landscape garden and will also bring unprecedented surprises to the public and to Chinese and foreign tourists. Architectural lighting and nightscape photography transcend architecture itself, and can transcend all fields of landscape gardening. Since the pictures presented in this book are recreations of existent ancient landscape gardens, they offer a new appreciation focus through a photographer's unique perspective. The impact of the images manifests the buildings' humanistic connotation and their precious heritage value. The art of nightscape lighting in Summer Palace is supported not only by hardware, but also by the improvement of soft power. The images of architectural photography have integrated the ancient landscape gardens with the art of photography. Below please follow us into the nightscape of Summer Palace to feel its unique charm and elegant mood under the night sky.

1
文昌阁远眺

谐趣园
Garden of Harmonious Pleasures

云移溪树侵书幌
风送岩泉润墨池

一峰则太华千寻，一勺则江湖万里。山与水的景致因造园智慧而大放异彩，因凝缩天地而谱就新章。

In art of landscape gardening, a peak represents a mountain of a thousand xun (measurement of length), a spoonful of water, rivers and lakes. The scenery of the mountains and the waters demonstrates extraordinary wisdom of gardening art, because it creates a new world by condensing natural objects.

1
谐趣园景

1
光影走廊

自一段幽深狭窄的山路起始,绕过遮天蔽日的古树步入宫门,眼前豁然出现一潭碧水,这便是"步步皆景"的谐趣园。谐趣园坐落于万寿山东麓,是依照江南园林精妙、清奇风格设计的园中之园。因园子仿照无锡惠山寄畅园而建,曾名"惠山园"。嘉庆十六年(1811年)重修后,取"以物外之静趣,谐寸田之中和"和乾隆皇帝的诗句"一亭一径,足谐奇趣"的意思,改名为"谐趣园"。

Go for a stroll starting from a narrow, deep, and serene mountain path, bypass the towering ancient trees, and step into the gate of a palace. When suddenly greeted by a pool of clear water, you are in the Garden of Harmonious Pleasures, where you are presented at every step with new beautiful scenery. The Garden of Harmonious Pleasures, located at the eastern foot of Longevity Hill, is a garden within a garden designed in light of the sophisticated and eccentric style of the landscape gardens found in the area south of Yangtze River. Because the garden modeled on the Jichang Garden in Huishan, Wuxi, it was formerly known as "Huishan Garden". After the renovation in the 16th year of Jiaqing's reigh (1811), it was renamed as the "Garden of Harmonious Pleasures," taking the meaning that "With the quiet pleasures of enjoying worldly possessions, we achieve a harmonious balance of mind and spirit" and that "Even one pavilion, or one footpath, brings harmonious pleasures," a verse by Emperor Qianlong.

园内十三处建筑以廊、桥相连，池映廊景，化实为虚，构筑成一个回环往复的江南遗梦。暮色四合之时，廊桥变幻的灯光拂去历史的尘埃，照亮含烟的清池。古人在建筑彩绘中叙述着历史典故、民间传说，又在婉转迂回中增加了几分情趣。园内最著名的景致莫过于东南角一座石桥，桥头石坊上有乾隆题写的"知鱼桥"三字，是引用了庄子和惠子在"濠上"的争论而来的。

In the park of Summer Palace, the 13 buildings are connected with corridors and bridges and the virtual corridors reflected in water create a circle of lingering dreams of the sunny south. When the dusk is sinking, the changing light of the corridors and bridges sweeps away the dust of history, illuminating the clear, umber-black lakes. The ancients in the architectural paintings narrate the historical allusions and folk legends, adding inclinations and interests in tactful and roundabout ways. The most famous scene in the park is a stone bridge at the southeastern corner, and there are three characters, "Knowing Fish Bridge," inscribed by Emperor Qianlong on the stone arch at the bridge end, which quoted from the debate between Zhuangzi and Huizi on Hao River, an event recorded in "Zhuangzi: Autumn Water."

1
庭院深深

2
夜阑犹剪灯花弄

1
谐趣园连廊

石栏遮曲径,春水漾方塘。待天边的红霞渐退,池中光影愈加明亮,夜景灯光下的谐趣园笼罩在一种优雅清奇的氛围中,远看朦胧统一,近看细节丰富,移步前行各种景物强弱分明,节奏有序,精巧的布局更显其构成严谨、错落有致,细节处蕴藏着隽永的意味,为灯下美景带来巧夺天工的情趣。

Stone railings hide the winding paths; spring water brims over the square pond. While red evening glow on the horizon fades away, light and shadow in the lake become brighter. The Garden of Harmonious Pleasures, in the nightscape light, is shrouded in a graceful, silent atmosphere. It is misty but unified when viewing it from afar, and rich in details when seeing it closely. When you move forward, every scene becomes distinct in intensity, and has orderly rhythm. Its elaborate layout shows it was conscientiously and carefully constructed, and irregularly arranged with charming effects. The details contain thought-provoking overtone, and bring fun to the beautiful night scenery.

"知鱼桥"下,点点荷花灯映照碧水,伴着精美的园区悄然入梦,"子非鱼焉知鱼之乐"的典故犹在耳旁,不知鱼儿是否也喜欢这水中的明亮世界。

Under the Knowing Fish Bridge, dotted lotus lanterns are reflected in the green water, accompanying the beautiful park in a quietly dream. "You are not a fish. How do you know the fish is happy?" The classical story is still ringing in the air. I wonder if fish like the bright world in the water.

谐趣园

谐趣园照明主体是重点游览景观，属于一类文物建筑，有油饰彩画分布。与建筑接触的灯具采用小功率线性洗墙灯进行抱箍式安装，不破坏建筑本体。且灯具可随时拆除，拆除后不对建筑产生影响，具备可逆性。

廊道内彩画表面水平照度

计算面	平均值(lx)	最小值(lx)	最大值(lx)	最小值/平均值	最小值/最大值
	15	0.06	24	0.004	0.003

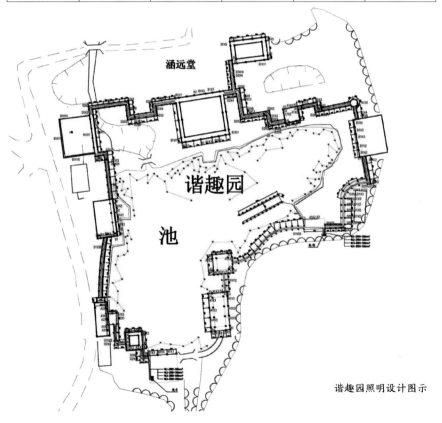

谐趣园照明设计图示

外廊彩画表面水平照度

彩画面	平均值(lx)	最小值(lx)	最大值(lx)	最小值/平均值	最小值/最大值
	13	3.94	25	0.303	0.158

照射时间：每日照明开启时间应小于 3 h。

照明光源：采用不含紫外成分的 RGB 型 LED 光源。

水下灯具本身较为隐蔽，不对景观产生影响。

灯具采用仿色以及仿生等手法，降低对景观的影响。

所使用灯具全部满足相关安全防火、防水等级要求。

选用高效节能型 LED 光源，同时在灯具与建筑结合面上进行热阻隔垫片处理，降低热辐射对建筑的影响。

采用 DMX512 主控系统，实现功能照明与装饰照明分场景、分时段进行控制；同时达到装饰照明实现 RGB 颜色变化及亮度 5%～100% 的调节。

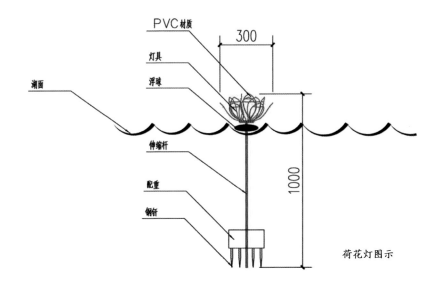

荷花灯图示

1
向北堪骋望,绿云迷万顷

霁清轩
Jiqing Pavilion

溪烟敛归岫
花露垂恋枝

霁清轩始建于乾隆十九年（1754年），这里原属于惠山园的一部分，嘉庆时期改造后独成一院。霁清轩的园林设计利用居高临下的山势，依次布置山谷、溪流、廊亭、松树等景观，并用一组连廊构成连续的空间元素，仿佛一条盘龙依山而卧，伺机腾空而起。

Jiqing Pavilion was built in the 19th year of Qianlong's reigh (1754). This place originally belonged to part of the Huishan Garden, but after Jiaqing era, it became an independent court. The landscape design of the pavilion takes advantage of descending hills, where the valleys, streams, pavilions, pines, etc. are arranged in an orderly way and connected by a set of corridors to form a continuous space element, as if a dragon was lying on the hill and waiting for the opportunity to soar.

进入垂花门,正对霁清轩。霁清轩是主殿,其特有的海墁绿地藤萝花彩画装饰,是该建筑最具特色之处。这在颐和园也是绝无仅有的。

After entering the gate with suspended pillars, you are directly facing the Jiqing Pavilion. Jiqing Pavilion is a main palace hall. Its unique color painting decoration of wisteria flower in green plaster background is the most characteristic of the building. This is also unique in Summer Palace.

1
霁清曲廊

2
霁清轩垂花门

霁清轩的西殿名清琴峡。乾隆皇帝曾赋诗道:"引水出石峡,抱之若清泉,峡即琴之桐,水即琴之弦。"此处造园者因山制宜,将后溪河的河水一分为二,一水向南,经玉琴峡流向谐趣园;一水向东,穿过地下人工暗道从清琴峡三间殿下汩汩而出,顺着凿成的峡谷流向圆明园。

The West Palace in Jiqing Pavilion is named Qingqin Gorge. Emperor Qianlong wrote a poem: "Channel water out the gorge, and hold it like clear spring. The gorge is paulownia of qin, and the water is string of qin." Landscapers appropriately used the hills to divide Houxi River's water into two streams. A stream flows southward, passing the Yuqin Gorge to the Garden of Harmonious Pleasures. The other stream flows eastward into a man-made underground channel and bubbles out from the Sanjian Palace of the Qingqin Gorge, passing through a dug canyon into the Yuanmingyuan Imperial Garden.

1
霁清轩园景

1
深苔攀云顶

2
窗灯照月明

清琴峡和玉琴峡的造型不同，音色也各有不同。玉琴峡曲折错落水声激荡，如打击乐器铿锵有力；清琴峡直通平缓，流水淙淙，像弦乐般委婉动听。灯光下的霁清轩，更显示出变幻莫测、玄妙魔幻的境界。

Qingqin Gorge and Yuqin Gorge are different in shape, and also in tone color. The water sound of the zigzag Yuqin Gorge is sonorous, and powerful, resembling percussion music instrument. Qingqin Gorge is straight and gentle, and its murmuring of running water is like string music, euphemistic and pleasant. The Jiqing Pavilion, in nightscape lighting, is mysterious, like a magical ream.

1
四方亭暮色

2
幽谷传清音

霁清轩的最高点是四方亭。登临亭中，可远眺乡野，周围田园景象尽收眼底。夜景中的四方亭尽显迷人之处，或凭栏小憩，对月冥思，发思古之幽情；或听琴观景，光影、琴音、水声交织在一起，似仙境般令人神往。

The highest point of Jiqing Pavilion is the Square Pavilion. Climb up into the pavilion, you can overlook the countryside and enjoy the panoramic views of the surrounding rural areas. The Square Pavilion in the night scene is extraordinary charming. You may lean against the railing for a rest and meditate nostalgia under the moon light, or listen to qin music and enjoy the views, while light, music, and sound of water playing around you, like in a fascinating fairyland.

霁清轩

霁清轩有油饰彩画分布，属一类文物建筑，系高安全消防等级，是重点游览景观。与建筑接触的灯具采用小功率线性洗墙灯进行抱箍式安装，不破坏建筑本体。灯具可随时拆除，拆除后不对建筑产生影响，具备可逆性。

廊道内彩画表面水平照度

计算面	平均值(lx)	最小值(lx)	最大值(lx)	最小值/平均值	最小值/最大值
	15	0.06	24	0.004	0.003

外廊彩画表面水平照度

彩画面	平均值(lx)	最小值(lx)	最大值(lx)	最小值/平均值	最小值/最大值
	14.2	4.21	23	0.296	0.183

照射时间：每日照明开启时间应小于 3 h。

照明光源：采用不含紫外成分的 RGB 型 LED 光源。

霁清轩照明方案及施工符合颐和园夜景照明对于彩画保护、施工保护、景观影响、安全防护的评估要求。

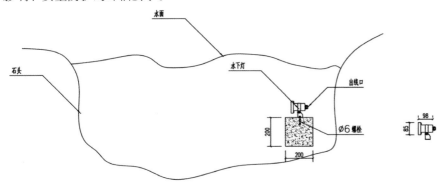

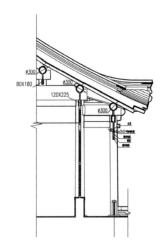
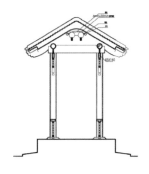

霁清轩灯光布局图

苏州街
Suzhou Street

北客南人成买卖
镜里功名待青罗

苏州街是后湖两岸仿江南水乡苏州而建的买卖街。建园早期，岸上有各式店铺，如玉器古玩店、绸缎店、点心铺、茶楼、金银首饰楼等，店铺中的店员都由太监、宫女装扮。皇帝游幸时开始"营业"。

Suzhou Street is a trading street on both sides of Houhu River and an imitation of water villages in Suzhou. In the early years of the park, there were various shops ashore, such as jade antique shops, satin shops, snack shops,

1
苏州街三孔石桥

restaurants, gold, silver and jewelry boutiques, etc., and the store clerks were court eunuches and maids, who dressed up as salespersons. The stores would "open for business", when the emperor was traveling the area.

1 苏州街夜景

后湖岸边的数十处店铺于 1860 年被焚毁，1986 年开始重建，1990 年竣工。建筑布局模仿了浙东一带常见的"一水两街"的形式，以后溪河中段的三孔石桥为中心向两侧展开。

Dozens of shops on the sides of Houhu River were burned in 1860. They were rebuilt in 1986 and completed in 1990, whose architectural layout imitated the one-river-two-street style found common in the eastern part of Zhejiang Province. The three-arch stone bridge in the middle of Houxi River is the center, from which the street spreads to both sides.

颐和园苏州街可以说是皇家园林的真人秀。帝王想体验市井生活又不便于经常微服出访，只好在自家园子里做了一个真实版的市场游戏，各种微缩后的商铺在灯光照射下显现出各具特色的格局和风貌。

Suzhou Street in Summer Palace can be said to be a reality show in royal gardens. Emperors wanted to experience the life in a marketplace, but it was not easy for them to visit the place often. Therefore, they had to make a real version of a marketplace to play in their own gardens. In the market, a variety of miniature shops in the lighting reflect the characteristics of the patterns and styles of different trade categories.

苏州街虽具有江南的水街市井特色，却沿用了北方店铺的式样，可见仿建结合场地条件和实际需要进行了再创作，捕捉到景致的创意与韵味。

Although Suzhou Street has the features of a southern city's waterfront downtown, its stores are in the northern style. Therefore, the imitation had integrated with the local conditions and actual needs in re-creation, in order to capture the creativity and lingering charm of the scenery.

1
披红挂绿的苏州街

2
苏州街小景

夜晚的苏州街灯火辉煌,两岸 60 余座店铺的灯光交织成一张梦幻的网,散发着绚烂的光辉。夜凉如水,湖面的微风吹动着灯笼、招幌在空中摇曳,更显江南水乡的韵味。夜幕下的苏州街在后湖灯光的映衬下风姿更甚,灯光、月光、水光的交相辉映,让人真的有置身江南水乡之感。

At night, the lights of Suzhou Street are dim, and the lights of more than 60 shops on both sides of the river are woven into a dream net, which emits

1
苏州街三孔桥以西全貌

2
精美雕饰还原江南遗梦

splendid brilliance. The night is as cool as water, and the lanterns and posters sway in the lake breeze, showing the lingering charm of the water village in Jiangnan. At night, the graceful bearing of Suzhou Street, against the lights from Houhu River, becomes even more enchanting, with lights, moonlight, and water reflections adding radiance and beauty to each other, making people feel as if they were placing themselves in the water village in Jiangnan.

1
璀璨迷幻的澹宁堂

苏州街

苏州街主体有油饰彩画，同样是一类文物建筑，重点景观。为保护珍贵的彩画艺术，多采用 6W、12W 的小功率 LED 光源，且照明非直射彩画，而是利用逸散光为彩画提供照明，节省能耗的同时大大降低了人工光环境对建筑的损伤。灯具采用抱箍式非破坏可逆性方式安装。将灯具安装于檐口下方等隐蔽位置，利用建筑自身掩蔽灯具，不对建筑景观产生影响。安全性方面使用符合安全要求的灯具及线缆，同时使用隔热垫片避免灯具发热对建筑产生的影响。

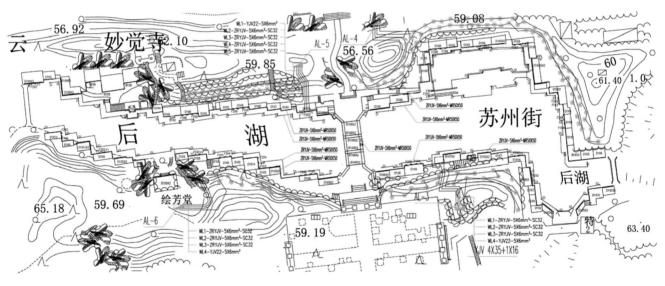

苏州街布灯平面图

廊道内彩画表面水平照度

彩画面	平均值（lx）	最小值（lx）	最大值（lx）	最小值/平均值	最小值/最大值
	4.65	0.16	27	0.034	0.006

照射时间：每日照明开启时间应小于 2 h。

照明光源：采用不含紫外成分的 RGB 型 LED 光源。

流光溢彩的苏州街

后溪河
Houxi River

坐看云起堆石处
吟诗摇橹后溪河

后湖位于颐和园最北部,又名后溪河,全长约1200米,河道蜿蜒于万寿山北麓山坡。后溪河由人工开挖而成,挖出的土全部堆放在河的北岸,形成土山,与万寿山隔河相望,并形成两山夹一河的格局。河面宽窄相间,形成大小不同的6个小湖,极具江南风韵。后溪河西侧在界湖桥与昆明湖相连,东南至紫气东来城关附近流出宫墙汇入畅春园。

光幻湖山——颐和园夜景灯光艺术鉴赏

1
后溪河夜景

1
迷幻的灯光照亮绮望轩遗址

2
双桥的灯带打造出鹰目般的效果

Houhu Lake is located in the northernmost part of Summer Palace, also known as Houxi River, with a full length of about 1,200 meters, winding in the north foot of Longevity Hill. The river is made by manual excavation. The soil dug up was all piled on the north bank of the river and formed an earth hill, standing opposite to the Longevity Hill across the river, becoming a pattern of one river between two hills. The river is alternately wide and narrow, forming 6 small lakes of different sizes, with great Jiangnan's charm. Houxi River connects Kunming Lake at Lake Dividing Bridge, and flows southeastward to the neighborhood of Ziqi Donglai Gate and out from the walls and then into Changchun Garden.

绮望轩、看云起时等虽是遗址，但从其精美的石制蹬道，可看出原建筑的精巧别致。复建的澹宁堂隐秘宁静，半璧桥的精心雕凿，林彪桥的历史传说，再加上河两岸绿水青山，林木葱茏，山路曲折，幽雅恬静，使其成为颐和园内清幽静谧之所在。这里两山夹一河的景致，与昆明湖西堤两水夹一堤的景色形成鲜明的对比。

Qiwang Pavilion and the site of Looking at Rising Clouds, among others, are ruins, but their exquisite stone pedaling roads can reveal the fineness of the original architecture. The secludedness and tranquility of the reconstructed Danning Hall, the careful carving of the Half Bridge, and the historical legend of Lin Biao Bridge, coupled with the green hills with verdant and thick foliage, the twists and turns of the hill paths, are elegant and silent, so that the area becomes a quiet and beautiful place in Summer Palace. The landscape of one river between two hills is in sharp contrast to the landscape of one hill between two rivers of the West Dike of Kunming Lake.

1
冷暖色的冲击赋予了半壁桥灵魂

2
拱券之曲

光幻湖山——颐和园夜景灯光艺术鉴赏

夜景环境中的后溪河有很多魔幻的灯光效果，微风轻摇着树枝，微澜的河面发生了轻妙的变化，水中月破碎又重圆；园子里夜行小动物的出没触发了一些巧妙的机关，灯光倏明倏暗，给宁静的后溪河赋予了一个跃动的灵魂。半壁桥洞与水面倒影恰巧构成一块完整的玉璧，灯光掩映下宛如一轮明月。夜游后溪河，好似踏上去往秘境的旅程。

In the nightscape, Houxi River has many magical lighting effects. The breeze gently sways the branches, and makes gentle and marvelous changes in the slightly waving surface of the river, where the moon is broken and reunited again. Some nocturnal little animals in the garden triggere some ingenious gears, and the lights suddenly darkenand suddenly brighten, giving a vibrant soul to the tranquil river of Houxi River. The arch of Half Bridge and its reflection in water happen to constitute a complete jade annulus, which resembles a bright full moon with the backdrop of lighting. A night trip to Houxi River is like a journey to a secret realm.

后溪河

后溪河的照明主体是水面及沿岸的古树群,为展现其风流韶华、清婉绮丽之貌,需要璀璨的灯光点亮园区,用光环境艺术尽展水生态、水文化、水人文之魅力。

后溪河水面水平照度

水面	平均值(lx)	最小值(lx)	最大值(lx)	最小值/平均值	最小值/最大值
	10	0.32	109	0.032	0.003

草地表面水平照度

水面	平均值(lx)	最小值(lx)	最大值(lx)	最小值/平均值	最小值/最大值
	0.62	0.000	7.89	0.000	0.000

照射时间:每日照明开启时间应小于 2 h。

照明光源:采用不含紫外成分的 RGB 型 LED 光源,最大限度避免了古迹遭受光污染。

在布局中特别注意灯具设备对环境风貌的影响,安装于码头的灯杆具有标志性作用,作为景观的一部分,不会对景观产生影响。位于沿岸道路边缘的灯具,安装于较为隐蔽的位置,也不会对景观产生影响。灯具采用仿色以及仿生等手法,降低对景观的影响。

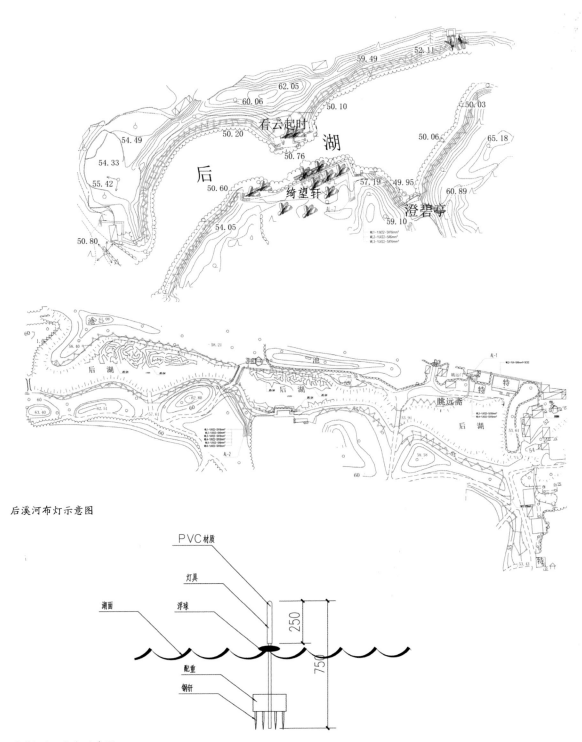

后溪河布灯示意图

后溪河水下灯杆示意图

1
美轮美奂的长廊

长廊
Long Corridor

邀月同赏长廊画
山色湖光共一楼

长廊是世界上最长的画廊兼游廊,建成于公元 1750 年,位于昆明湖与万寿山之间,长廊不仅连接了山水,更为全园增添了神趣,可谓巧夺天工之作。

Long Corridor is the world's longest gallery and corridor, built in 1750, located between Kunming Lake and Longevity Hill. It not only connects the hills and waters, but also adds mysterious interests to the entire park, and can be considered a masterpiece.

1
夜幕难掩长廊的风姿

长廊东起邀月门，西至石丈亭，长廊全长 728 米，分为 273 间，中间点缀着留佳、寄澜、秋水、清遥四座重檐八角攒尖亭，象征春夏秋冬四季。

Long Corridor starts from the east end, Yaoyue Gate, and extends to the west end, Shizhang Pavilion. The corridor is 728 meters long and is divided into 273 bays. In the middle, the corridor is dotted with four octagonal pavilions, each with double-leveled eaves and a pyramidal roof, named Liujia, Jilan, Qiushui, and Qingyao, symbolizing the four seasons, spring, summer, autumn, and winter, respectively.

灯光照亮了彩画里不为人知的细节，也给这个"中国古代彩绘博物馆"营造出美轮美奂的意境。长廊彩画是苏式彩画最具代表性的作品，绘有彩画14000余幅，这些彩画大体上可分为人物、山水、花鸟以及中国四大名著（《红楼梦》《三国演义》《西游记》《水浒传》）中的情节。许多风景画仿自江南水乡，是画师们根据乾隆皇帝的意图绘制的。

1
长廊夜色

2
令人惊叹的彩画艺术

Lights illuminate the unknown details of the paintings and also create a magnificent mood for this ancient Chinese colorful painting museum. There are more than 14,000 colorful paintings displayed in the corridor and they are the most representative works of the Su-style colorful paintings, including, roughly, four types, such as figure paintings, landscape paintings, flower-and-bird paintings, as well as plot paintings in four famous Chinese classics (*The Dream of the Red Chamber, The Three Kingdoms, Journey to the West, and The Water Margin*). Many landscape paintings imitate the landscape of the water villages in Jiangnan, drawn by the artists based on the intention of Emperor Qianlong.

夜景中的长廊横贯万寿山山麓，恰似沿昆明湖北岸东西逶迤的一条"光带"，游人置身廊间漫步，便会油然产生一种"人在廊中走，神在画中游"的感觉。抬眼望去，远处的西堤、湖心岛、十七孔桥尽收眼底，皇家园林特有的画意诗情让人难以忘怀。

At night, the Long Corridor traverses the foothills of Longevity Hill, just as an east-west "strip of lights" along the northern shore of Kunming Lake. Visitors, strolling down the corridor, will naturally have a kind of feeling that when walking in the corridor, his/her mind is travelling in each painting. Look up, and you will have a panoramic view of the distant West Dike, the Mid-lake Island, and the Seventeen-arch Bridge. The unique idyllic conception of the royal garden is unforgettable.

1
秋水亭夜色

光幻湖山——颐和园夜景灯光艺术鉴赏

长廊

长廊属一类文物建筑,其古建筑表面有大量丰富的油饰彩画,因此消防安全级别很高。

长廊保护性照明控制指标

等效照量			灯具相关			光源	
照明主体表面最高照度(lx)	照射时间(h)	每天等效照量(lx·h)	遮光限制	灯具安装限定	灯具外形、位置限制及日间处理措施(风貌)	安全等级	色温(K)	光谱限制(紫外线、红外线)
200	<3	<500	不直射彩画	禁止破坏文物的灯具安装	灯具安装位置隐蔽,采用仿色等手段与周围环境协调	Ip 65	3000	不含有紫外线的低温光源

廊道内彩画表面水平照度

彩画面	平均值(lx)	最小值(lx)	最大值(lx)	最小值/平均值	最小值/最大值
	15	0.06	24	0.004	0.003

照射时间:每日照明开启时间应小于 3 h。

照明光源:采用不含紫外成分的 RGB 型 LED 光源。

灯具设备对环境风貌的影响评估:所采用的线性洗墙灯尺寸较小,利于进行掩蔽。建筑照明中将灯具安装于顶部结构中,利用建筑自身遮挡灯具,不对景观产生影响。

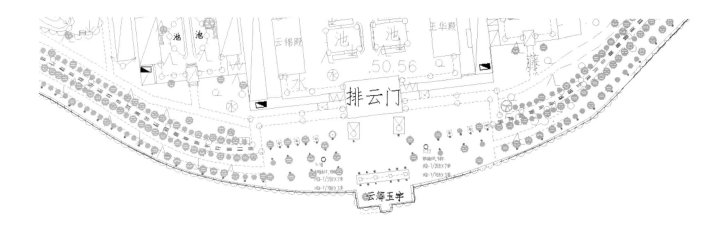

万寿山·佛香阁
Longevity Hill and Tower of Buddhist Incense

高峰称万寿
慈寿祝同齐

万寿山属燕山余脉,高 58.59 米,万寿山前山以八面三层四重檐的佛香阁为中心,组成巨大的主体建筑群。建筑群依山而筑。从山脚的"云辉玉宇"牌楼,经排云门、二宫门、排云殿、德辉殿、佛香阁,直至山顶的智慧海,形成了一条层层上升的中轴线。

1
万寿山建筑群

2
佛香阁细部

1
烟雨迷蒙的佛香阁

Longevity Hill belongs to the Yanshan mountains, with 58.59 meters high. The building complex in front of Longevity Hill is centered on the Tower of Buddhist Incense, an eight-sided, three-storied pavilion with four-leveled eaves. The buildings were constructed along the hillside. From the Yunhui Yuyu Archway at the foot of the hill, to Paiyun Gate, Ergong Gate, Paiyun Hall, Dehui Palace, Tower of Buddhist Incense, and the Sea of Wisdom on the hilltop, forms a layered rising central axis.

佛香阁位于万寿山前山中央部位的山腰，是颐和园建筑的标志，也是颐和园建筑布局的中心。佛香阁建筑在一个高21米的方形台基上；阁高41米，阁内有8根巨大铁梨木擎天柱。原阁于咸丰十年（1860年）被英法联军烧毁，后于光绪十七年（1891年）重建，光绪二十年（1894年）竣工，是颐和园里最大的工程。阁内供奉着"接引佛"，供皇室在此烧香。

The Tower of Buddhist Incense is located in the central hillside in front of Longevity Hill, and is a symbol of Summer Palace architecture and the center of Summer Palace's building layout. The tower was built on a 21-meter-high square foundation, and is 41 meters high, with 8 huge lignumvitae wood as suppor pillars. The original tower was destroyed by the Anglo-French Allied Force in the 10th year of Xianfeng's reign (1860), reconstructed in the 17th year of Guangxu's reign (1891) and completed in the 20th year of Guangxu's reign (1894), which was summer Palace's largest building project. Ambassador Buddha was enshrined in the tower, for the royal family to burn incense.

作为南北中轴线上最为宏大的建筑，佛香阁在夜景灯光中尽显宏伟气势，弥散的梵音萦绕在绿云一样的山麓之中，形成众星捧月般的气势。在周围夜景的衬托下，这百年间的光影更显庄重威严。

As the most grandiose structure on the north-south central axis, the Tower of Buddhist Incense has been magnificent in the nightscape lighting, with dissipated Buddhist sounds lingering in the green-cloud- like foothills, forming the momentum of a host of lesser lights around the leading one. Against the surrounding night scenes, the century-old building's light and shadow are more dignified and majestic.

1
巍峨的万寿山中轴线建筑群

排云门

排云门是进入佛香阁景区游览的主要入口，门前较为开阔，客流相对较大，照明从以下方面考虑：照明系统应弱化，减少照明设备对环境的影响；照明设备与现场相关设施相结合，弱化灯具对景观的影响；照明设备外表面颜色要与周围环境相协调。

排云门彩画表面水平照度

彩画面	平均值(lx)	最小值(lx)	最大值(lx)	最小值/平均值	最小值/最大值
	117	20.5	188	0.175	0.109

照射时间：每日照明开启时间应小于 3 h。

照明光源：采用不含紫外成分的 RGB 型 LED 光源。

排云门布灯示意

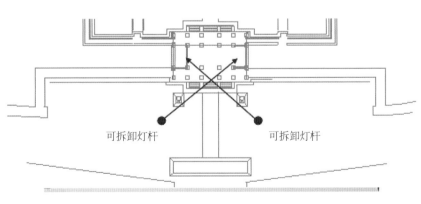

可拆卸灯杆　　可拆卸灯杆

二宫门

通过排云门进入二宫门,二宫门属重点景观,院墙外有古树,建筑有油饰彩画。

二宫门彩画表面水平照度

彩画面	平均值(lx)	最小值(lx)	最大值(lx)	最小值/平均值	最小值/最大值
	103	15.2	174	0.148	0.087

照射时间:每日照明开启时间应小于3 h。

照明光源:采用不含紫外成分的RGB型LED光源。

灯具采用投光灯,使用灯杆安装,不破坏建筑本体。

灯具、灯杆可随时拆除,拆除后不对建筑产生影响,具备可逆性。

灯具设备对环境风貌的影响评估:投光照明灯杆院墙外侧灯安装在山体土石地表处,颜色与周围整体色调相协调,降低对环境的影响。

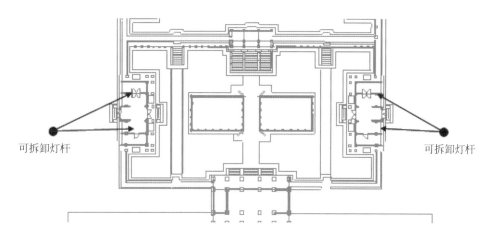

二宫门布灯示意

排云殿

灯具采用投光灯，使用灯杆安装，不破坏建筑本体。

灯具、灯杆可随时拆除，拆除后不对建筑产生影响，具备可逆性。

灯具设备对环境风貌的影响评估：投光照明灯杆院墙外侧灯安装在山体土石地表处，颜色与周围整体色调相协调，降低对环境的影响。

所使用灯具全部满足相关安全防火等级要求。

选用高效节能型 LED 光源，同时在灯具与建筑结合面上进行热阻隔垫片处理，降低热辐射对建筑的影响。

采用 DMX512 主控系统，实现功能照明与装饰照明分场景、分时段进行控制；同时达到装饰照明实现 RGB 颜色变化及亮度 5%～100% 的调节。

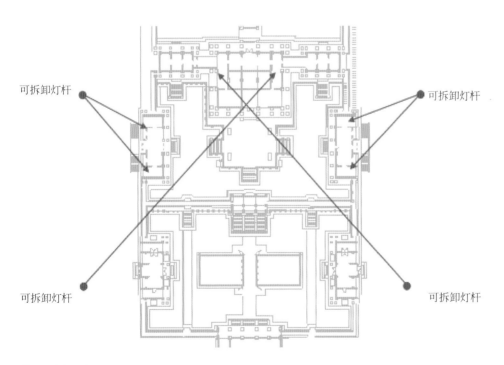

排云殿布灯示意

排云殿保护性照明指标控制

等效照量			灯具相关				光源	
照明主体表面最高照度（lx）	照射时间（h）	每天等效照量（lx·h）	遮光角限制	灯具安装限定	灯具外形、位置限制及日间处理措施（风貌）	安全等级	色温（K）	光谱限制（紫外线、红外线）
200	<3	<500	不直射彩画	禁止安装在建筑物及附近构筑物上	灯杆立于院墙外开敞处，做成升降杆或者倾倒杆	Ip 65	3000	不含有紫外线的低温光源

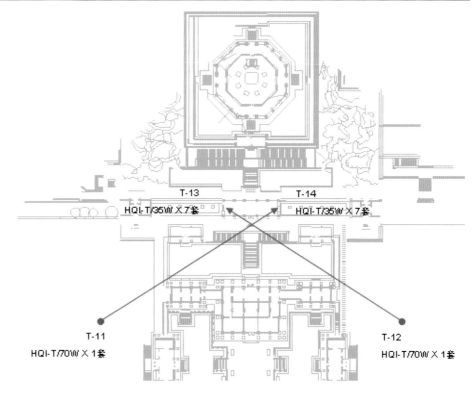

德辉殿布灯示意

佛香阁

佛香阁属于一类文物建筑，有油饰彩画分布，视野开敞，为重点游览景观。其照明灯杆布置方案如下图所示。

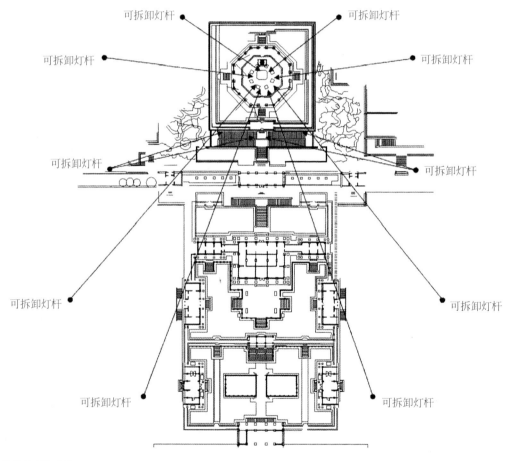

佛香阁布灯示意

远眺佛香阁

其投光照明灯杆合理安放在院落外部山体上，灯具、灯杆均耐腐蚀。选用可升降灯杆及可拆卸灯杆，且全部做成假树景观，弱化照明设施对园林景观建筑风貌的影响，保护万寿山景观风貌。

佛香阁照明方案及施工符合颐和园夜景照明对于彩画保护、施工保护、景观影响、安全防护的评估要求。

西堤
West Dike

琳琅三竺宇
花柳六桥堤

西堤位于昆明湖上，自西北逶迤向南。西堤及其支堤把湖面划分为三个大小不等的水域，每个水域各有一个湖心岛，象征中国古老传说中的东海三仙山——蓬莱、方丈、瀛洲。

The West Dike is located on Kunming Lake, meandering from northwest to south. The West Dike and its branch dikes divide the lake into three water bodies of unequal sizes, each of which has a mid-lake island, symbolizing the ancient Chinese legend of the Three East Sacred Mountains—Penglai, Fangzhang, and Yingzhou.

1
西堤桥亭

西堤一带碧波垂柳,自然景色开阔,园外数里的玉泉山秀丽山形和山顶的玉峰塔,构成了园景的组成部分。从昆明湖上及东堤西望,园外之景和园内湖山浑然一体,这是中国园林中运用借景手法的杰出范例。

1
西堤全景

In the West Dike area, there are blue waves and weeping willows, and the natural scenery is open. The beautiful shape of Yuquan Mountain and the Yufeng Pagoda on the top of the mountain, several li away from the park, are considered as elements of the landscape. Viewed westward from Kunming Lake and the East Dike, the scenery outside the park and the lakes and hills in the park are seamlessly connected into a whole. This is an outstanding example of view borrowing technique in Chinese landscape gardening.

西堤全长 2.2 公里。与杭州西湖的苏堤相似,纵贯颐和园昆明湖南北的西堤上,也建有六座造型各异的桥梁,合称西堤六桥。从北向南依次是:界湖桥、豳风桥、玉带桥、镜桥、练桥、柳桥。练桥和柳桥中间建有景明楼。

West Dike is 2.2 km in length. Similar to West Lake in Hangzhou, there are also six bridges of different shapes on the West Dike, which runs through Summer Palace and Kunming Lake, collectively known as Six Bridges on the West Dike. From north to south, they are Lake Dividing Bridge, Binfeng Bridge, Jade Belt Bridge, Mirror Bridge, White Silk Bridge. and Willow Bridge. Between the White Silk Bridge and the Willow Bridge, there is a building named Jingming Building.

1 2 3
多种光线下的景明楼

1
豳风桥

2
界湖桥

如果说长廊像交响乐的展开部，灯光下的西堤更像是一段舒缓的间奏。可以想象耳边响着舒缓的音乐，极目远望，灯光在西山的背景下低调而舒缓地律动着，更显湖水的宁静与安详；在灯光色彩变化过程中，可看到光影变化的奇妙，如同欣赏一首动听的音诗；西堤上六座桥则是间奏中跳跃的音符,不断变化的色彩如同伴奏的和弦,映衬出佛香阁舒缓统一的灯光主题。

If the Long Corridor is like the development of the symphony, the light of the West Dike is more like a soothing interlude. We can imagine, while listening to the sound of soothing music and viewing into the distance, lights against the background of the West Hill move rhythmically and soothingly in a low key, and the serenity and tranquility of the lake become more apparent. In the process of changing light colors, we witness the magic of the changes, as if we were appreciating a moving poem. The six bridges on the West Dike are the leaping notes of the intermezzo, changing colors like the changing chords, reflecting soothing, unified theme of the lighting at the Tower of Buddhist Incense.

天色渐晚,灯光亮起,原本就安静的西堤更显静谧。同一片湖水,能够营造出不同的意境,古人设计之精良可见一斑。

It is getting late, the lights are on, and the originally quiet West Dike is now more peaceful. The same lake can create different artistic conceptions; therefore, the excellence of ancient design is obvious.

1
景明楼

2
西堤湖畔

1

2

1
玉带桥

2
镜桥

3
精美绝伦的彩绘艺术

西堤六桥

西堤六桥属一类文物建筑，重点游览景观，无油饰彩画。

与桥体接触的灯具采用小功率线性洗墙灯进行抱箍式安装，且抱箍采用非金属材料，不破坏桥体本体。灯具可随时拆除，拆除后不对建筑产生影响，具备可逆性。

灯具设备对环境风貌的影响较小，其中所采用的线性洗墙灯尺寸较小，利于掩蔽；桥体照明中将灯具安装于桥洞下方，利用桥体自身遮挡灯具，不对景观产生影响；桥体立面所使用的投光灯安装于周围环境较为隐蔽的地方，不对景观产生影响；对照明中所使用的灯具采取仿色处理，降低对景观的影响。

所使用灯具全部满足相关安全防水等级要求。

选用高效节能型 LED 光源，节能环保。同时在灯具与建筑结合面上进行热阻隔垫片处理，降低热辐射对建筑的影响。

采用 DMX512 主控系统，实现功能照明与装饰照明分场景、分时段进行控制；同时达到装饰照明实现 RGB 颜色变化及亮度 5%～100% 的调节。

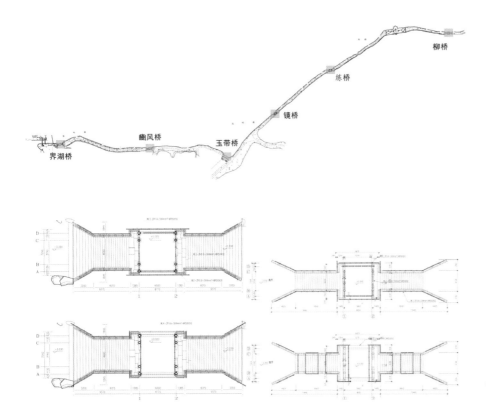

西堤六桥布灯示意及灯具示意

景明楼

景明楼属一类文物建筑，重点游览景观，有油饰彩画。

灯具可随时拆除，拆除后不对建筑产生影响，具备可逆性。

灯具设备对环境风貌的影响较小，所采用的线性洗墙灯尺寸较小，利于掩蔽；

将灯具安装于建筑檐口下方，利用建筑自身遮挡灯具，不对景观产生影响；

照亮山墙的远处投光灯放置在较为隐蔽的位置。

所使用灯具全部满足相关安全防火等级要求。

选用高效节能型 LED 光源，节能环保，同时在灯具与建筑结合面上进行热阻隔垫片处理，降低热辐射对建筑的影响。

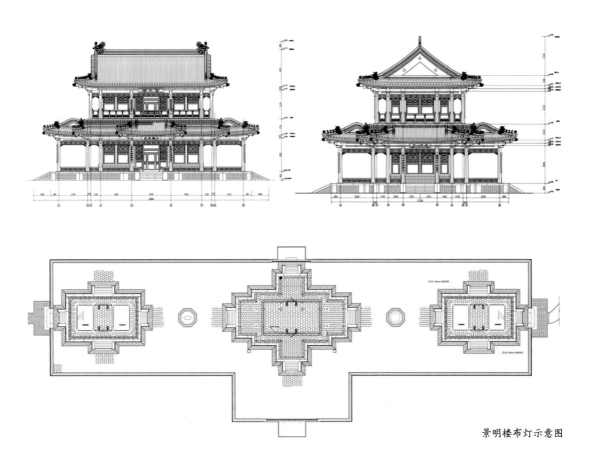

景明楼布灯示意图

1

九道弯·知春亭·文昌阁

Nine Bay, Zhichun Pavilion, and Wenchang Tower

遥其塞上翻林叠
近此园中印水潆

1
九道弯与佛香阁遥相呼应

2
九道弯

九道弯的夜景别有意境，围栏在灯光中富有韵律，沿湖灯光带与花窗中的灯光倒映于湖中，远处的佛香阁在湖水中留下金色的倒影，再加上湖中的荷花莲叶，可谓是移步异景，景随人动，令人流连忘返。

The night scene of the Nine Bay has a different kind of artistic conception. The railings have rich rhythm in the lights. The light ribbon and the lights in the windows are reflected in the lake. The distant Tower of Buddhist Incense leaves a golden reflection in the lake. In addition, there are lotuses and green leaves in the lake. When you move each step, you will find a different scene, and the scene moves, when people move, which is lingering and unforgettable.

1
灯光照亮九道弯

2
九道弯小品

3
九道弯建筑群

文昌阁在颐和园昆明湖东堤北端。原是一座城关，为清漪园的园门之一。建于清乾隆十五年（1750年），现存城楼为光绪时（1875—1908年）重建。城头四隅角廊平面呈"人"字形，中间为三层楼阁。中层供奉文昌帝君铜铸像及仙童塑像，旁有铜特。

The Wenchang Tower is located at the north end of the East Dike of Kunmimg Lake in Summer Palace. Originally, it was a wall gate. It is a moving picture at the gate of the Qingyi Yuan. It was built in the 15th year of Qianlong's reign (1750). The existing tower was the one reconstructed in Guangxu's reign (1875—1908). Corner corridors at the four corners of the wall are "人" shaped in the plane, with a three-storied pavilion in the middle. Wenchang Emperor's bronze statue is enshrined on the middle floor, accompanied by fairy child statues, and copper mules.

1
文昌阁

知春亭位于昆明湖东岸前高出湖面少许的人工堆筑的小岛上，是一座双围柱重檐四角亭。这里遍植桃柳，每当春天来临，此处最先柳绿花红，向人们报告春天到来的讯息，故名知春亭。

Zhichun Pavilion, located at a protruding little artificial islet on the east shore of Kunming Lake, is a square pavilion with two columns and double-leveled eaves. Peaches and willows are planted all over here, whenever spring comes, here is the first place to present people the green willows and red flowers; therefore, it is known as Zhichun Pavilion.

1 知春亭

文昌阁

文昌阁属一类文物建筑，有油饰彩画分布，是重点游览景观。

灯具采用小功率泛光灯通过柔性石棉垫与建筑接触，不破坏建筑本体。

灯具可随时拆除，拆除后不对建筑产生影响，具备可逆性。

灯具设备对环境风貌的影响评估：所采用的泛光灯尺寸较小，利于掩蔽。

建筑照明中将灯具安装于建筑门洞上方，利用建筑自身遮挡灯具，不对景观产生影响。

所使用灯具全部满足相关安全防火等级要求。

选用高效节能型LED光源，同时在灯具与建筑结合面上进行热阻隔垫片处理，降低热辐射对建筑的影响。

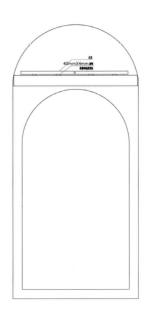

文昌阁保护性照明控制指标

等效照量			灯具相关				光源	
照明主体表面最高照度（lx）	照射时间（h）	每天等效照量（lx·h）	遮光角限制	灯具安装限定	灯具外形、位置限制及日间处理措施（风貌）	安全等级	色温（K）	光谱限制（紫外线、红外线）
200	<3	<500	不直射彩画	禁止破坏文物的灯具安装	灯具安装位置隐蔽，采用仿色等手段与周围环境协调	Ip 65	3000	不含有紫外线的低温光源

知春亭

知春亭属一类文物建筑,有油饰彩画分布,是重点游览景观。

知春亭照明方案及施工符合颐和园夜景照明对于彩画保护、施工保护、景观影响、安全防护的评估要求。

知春亭保护性照明控制指标

等效照量			灯具相关				光源	
照明主体表面最高照度（lx）	照射时间（h）	每天等效照量（lx·h）	遮光角限制	灯具安装限定	灯具外形、位置限制及日间处理措施（风貌）	安全等级	色温（K）	光谱限制（紫外线、红外线）
200	<3	<500	不直射彩画	禁止破坏文物的灯具安装	灯具安装位置隐蔽，采用仿色等手段与周围环境协调	Ip 65	3000	不含有紫外线的低温光源

1

1
彩光下的十七孔桥

十七孔桥·南湖岛·廊如亭

Seventeen-Arch Bridge, Nanhu Island, and Kuoru Pavilion

绣帘高卷倾城出
灯前潋滟横波溢

十七孔桥坐落在昆明湖上，位于东堤和南湖岛之间，用以连接堤岛，为园中最大石桥。石桥宽 8 米，长 150 米，由 17 个桥洞组成。石桥两边栏杆上雕有大小不同、形态各异的石狮 500 多只。

1
铜牛

2
夕阳下的十七孔桥

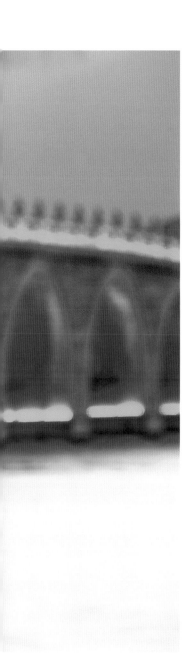

The Seventeen-arch Bridge is located on Kunming Lake, between the East Dike and South Lake Island, connects the island and the dike, and is the largest stone bridge in the park. The stone bridge is 8 meters wide and 150 meters long, and consists of 17 arches. The railings on both sides of the bridge have more than 500 carved stone lions with different sizes and shapes.

铜牛在昆明湖东岸，十七孔桥东桥头北侧，为镇压水患而设。1755年用铜铸造，称为"金牛"。

Bronze Bull on the east shore of Kunming Lake, to the north of the Seventeen-arch Bridge's east end, was built in order to suppress floods. It was made of copper in 1755, and called Golden Bull.

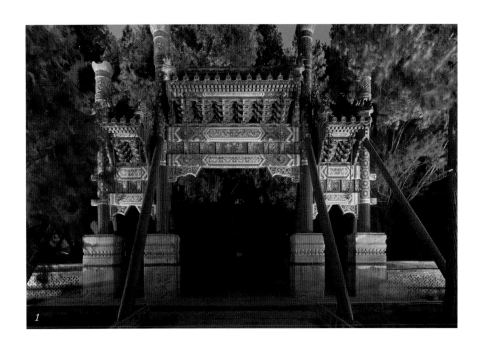

1
凌霄牌楼

2 3
廓如亭

廓如亭又名八方亭，它位于颐和园景区的东南，十七孔桥东端，与万寿山佛香阁隔湖相望，是颐和园内体量最大的一座重檐八角亭。

Kuoru Pavilion, also known as the Octagonal Pavilion, is located in the southeast of Summer Palace Scenic Area, at the eastern end of the Seventeen-arch Bridge, and standing opposite to the Tower of Buddhist Incense on Longevity Hill across the lake. It is an octagonal pavilion with double-leveled eaves and a large body mass in Summer Palace park.

1
十七孔桥与佛香阁同框

2
十七孔桥与廊如亭同框

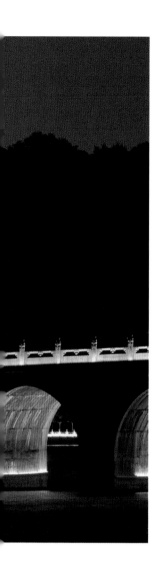

南湖岛位于昆明湖东南侧、万寿山南侧湖水之中，与万寿山遥相呼应，起到了丰富昆明湖水面景色的重要作用。

South Lake Island is located in the east of Kunming Lake, to the south of the Longevity Hill and opposite to Longevity Hill, and plays an important role of enriching the scenery of Kunming Lake.

与南湖岛和廓如亭相比，十七孔桥的夜景照明尤显突出，灯光笼罩下的十七孔桥与众星捧月般的佛香阁隔湖相望，佛香阁坐于万寿山上，庄重宏伟，十七孔桥卧于昆明湖上，秀丽婀娜，将十七孔桥和佛香阁置于同一框中，产生别样韵味。五彩斑斓的光影中，十七孔桥倒映于昆明湖中，本桥与倒影合为一体，夜色中的昆明湖令人陶醉。

1
月华西下露华凝

Compared with the South Lake Island, the nightscape lighting of the Seventeen-arch Bridge is especially outstanding. Under the lights, the Seventeen-arch Bridge faces the Tower of Buddhist Incense surrounded by attractions across the lake. The Tower of Buddhist Incense sits on Longevity Hall, solemn magnificent. and the Seventeen-arch Bridge is floating over Kunming Lake, beautiful and graceful. Putting the two in the same frame has a unique flavor. In multicolored light and shadow, the Seventeen-arch Bridge's reflection is in Kunming Lake, and the bridge is blended with its reflection as a whole. Kunming Lake in the night is intoxicating.

十七孔桥

十七孔桥属一类文物建筑，重点游览景观，无油饰彩画。

与桥体接触灯具采用小功率投光灯进行抱箍式安装，且抱箍采用非金属材料，不对桥体产生破坏。灯具可随时拆除，拆除后不对建筑产生影响，具备可逆性。灯具设备对环境风貌的影响较小，所采用的小型投光灯利于掩蔽；桥体照明中将灯具安装于桥洞下方，利用桥体自身遮挡灯具；桥体立面所使用的投光灯安装于周围环境较为隐蔽的地方，不对景观产生影响；同时对照明中所使用的灯具采取仿色处理，降低对景观的影响。

所使用灯具全部满足相关安全防水等级要求，选用高效节能型 LED 光源，同时在灯具与建筑结合面上进行热阻隔垫片处理，降低热辐射对建筑的影响。

十七孔桥保护性照明控制指标

灯具相关				光源	
遮光角限制	灯具安装限定	灯具外形、位置限制及日间处理措施（风貌）	安全等级	色温（K）	光谱限制（紫外线、红外线）
特制防炫光灯具	禁止安装在古建筑物上	灯具安装位置隐蔽，采用仿色等手段与周围环境协调	Ip 67	3000	不含有紫外线的低温光源

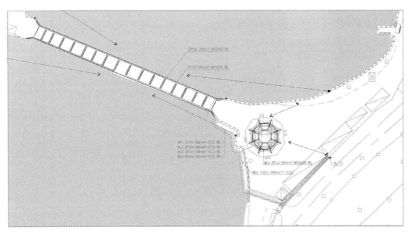
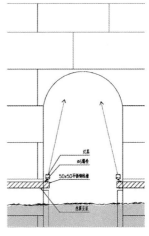

十七孔桥布灯示意图

南湖岛

南湖岛属一类文物建筑，重点游览景观，建筑有油饰彩画，其景观有观赏花木及少量古树。

建筑照明：采用小功率 LED 光源，照明不直射彩画，利用逸散光保护彩画。采用抱箍式非破坏可逆性安装方式。将灯具安装于檐口下方等隐蔽位置，利用建筑自身掩蔽灯具，不对建筑景观产生影响。使用符合安全要求的灯具及线缆，同时使用隔热垫片避免灯具发热对建筑产生的影响。

水体照明：采用节能的小功率 LED 光源，灯具安装于水下，利用水体掩蔽灯具。使用防水、耐腐蚀、符合安全防护要求的灯具及线缆。

南湖岛保护性照明控制指标

等效照量			灯具相关				光源	
照明主体表面最高照度（lx）	照射时间（h）	每天等效照量（lx·h）	遮光角限制（度）	灯具安装限定	灯具外形、位置限制及日间处理措施（风貌）	安全等级	色温（K）	光谱限制（紫外线、红外线）
200	<3	<500	不直射彩画	禁止安装在建筑物及附近构筑物上	灯具安装位置隐蔽，采用仿色等手段与周围环境协调	Ip 65	3000	不含有紫外线的低温光源

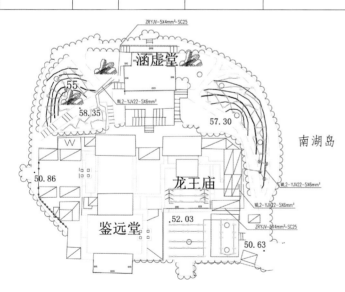

南湖岛布灯示意图

篇·三

颐和园夜景照明分析与启示
Summer Palace Landscape Lighting Analysis and Enlightenment

对建筑遗产保护之路，建筑学家林徽因有如此评说："知道一个民族在过去的时代里，有过丰富的成绩，不保证他们现在仍然在活跃繁荣的。但是反过来说，如果我们连祖宗留下来的家产都没有能力清理或保护，乃至于让家里的至宝毁坏散失……证明我们这做子孙的没有出息，智力德行已经到了不能堕落的田地。"本篇从颐和园夜景照明工程综合分析入手，不仅介绍了照明设计大纲及保护类照明设计要点，还探讨了针对全国"52处世遗"普适性的照明工程设计与施工管理"条例"的编研思路与建议，希望读者能从颐和园夜景照明工程设计实践中，感悟到有价值的技术策略与技术方法，从而推动中国文化遗产界照明技术的良好应用与发展。

Lin Huiyin, a Chinese architect, had ever made the remark below on architectural heritage protection, "If a nation made great achievements in the past, we shall not take for granted that the nation still enjoys prosperity and vitality at present. If we are incapable of sorting out or protecting the family properties left over by our forefathers, or even give way to damage and loss of our most valuable treasures, we have no prospects and our intelligence and moral conducts are on the verge of degradation." This chapter starts with a comprehensive analysis on the nightscape lighting project of Summer Palace, introduces the lighting design outline and the technical key points of protective lighting design, and puts forward thoughts and suggestions on the compilation of the "Rules" of a universal value for the lighting design and construction management of the 52 World Heritage Sites across China. We hope that the readers can acquire valuable technical strategies and methods from the design practices of Summer Palace nightscape lighting project, thereby promoting applications and development of lighting technologies for Chinese cultural heritages.

颐和园皇家园林保护工程

Program for the Preservation of Summer Palace Imperial Gardens

始建于1750年,中国最后一个封建盛世——"康乾盛世"期间的颐和园,集中呈现了中国皇家园林的富丽恢弘,处处可见、可感受"虽由人作,宛自天开"之意蕴。1860年第二次鸦片战争,颐和园(原名清漪园)被英法联军烧毁,1886年,清政府挪用海军军费等款项重修,后改为慈禧太后颐养之地,故正式改名为颐和园。至此,它成为除紫禁城之外晚清最高统治者的另一处

1
烟中列岫青无数

2
雁背夕阳红欲暮

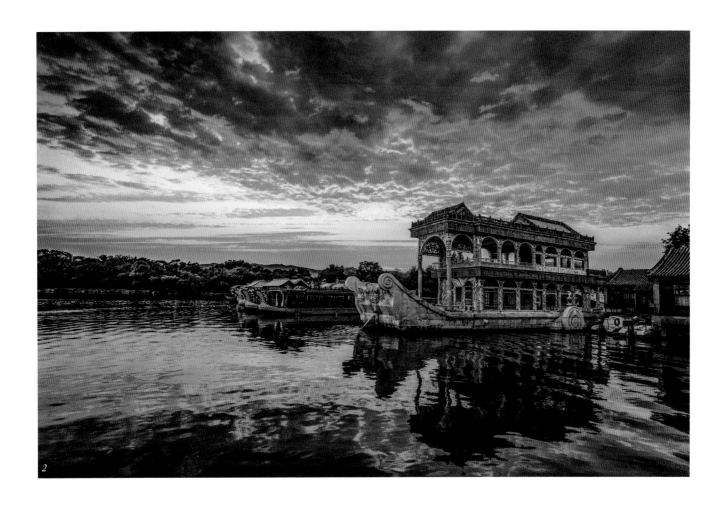

活动中心。清朝覆灭后，颐和园又曾作为溥仪私产对外开放，1928年被南京国民政府接管成为公园正式对外开放（2018年将迎来颐和园开放90周年）。中华人民共和国成立后，由于颐和园"北返文物"鉴定、提选和分配工作的完成，北京市公园管理委员会便开始布置对颐和园多处古建筑修缮的任务。

The construction of Summer Palace, initially called "Qingyi Garden "(Garden of Clear Ripples), began in 1750, in the last heyday of Chinese feudalism "Kangqian Era of Prosperity". It is a majestic and magnificent imperial garden, with its every part appearing as if "created by nature". In 1860, Summer Palace was burned down by the Anglo-French Allied Forces at the end of the Second Opium War. In 1886, it was reconstructed by the Qing government by using the fund that was originally designated for upgrading the Qing navy. Later, it became a place for Empress Dowager Cixi to lead a life of enjoyment and was given its present-day Chinese name. At that time, it became another activity center of the top ruler of Qing dynasty, apart from the Forbidden City. Summer Palace became the private property of Pu Yi, the last emperor of China and was opened to the public after the collapse of Qing dynasty. In 1928, it was taken over by the Nanjing National Government and opened officially to the public as a park (2018 will be the 90th anniversary of Summer Palace). As the work of authentication, selection and allocation of the cultural relics returned from south China had been done after the founding of PRC, Beijing Municipal Administration Committee of Parks began to renovate the attractions including Stone Boat at Summer Palace.

1953 年对佛香阁进行了清末以来第一次全面整修，倾斜 6 厘米的大木构架被重新拨正，全部阁顶重檐挑顶重修，补齐了琉璃瓦并描绘了金龙和玺彩画，贴大赤金。

1959 年，全长 728 米的长廊大修工程启动，也是自晚清以来长廊第一次全面修复，因采用苏式彩绘工艺，主要由工匠师们言传身教，1978 年前后，经多方招揽能工巧匠大展身手，才使许多模糊不清的彩画被绘制一新。

20 世纪 80 年代后至今，颐和园的修缮保护工程再掀高潮，如万寿山后山四大部洲、苏州街景观、昆明湖湖底清淤并重启开闸放水、西堤整治等。

2005 年 3 月 31 日，作为北京奥运会"人文奥运"文保项目重点工程之一的佛香阁景区修缮工程启动，它是中华人民共和国成立后颐和园规模最大的修缮工程。它严格采用传统工艺、传统材料，修补添配受损和缺失的构件，修补加固所有建筑的内外檐装饰,也对彩画中不符合历史原貌的部分重新绘制修整。

之所以用简洁的古建园林"修缮史"的方法盘点颐和园保护历程,是为了在颐和园这历史文化宝库和自然环境宝库中,展示当代颐和园逐渐完善的照明艺术,发现技术时代下的文化魅力,挖掘照明技术蕴含的文化精髓,感悟建筑摄影穿越并记忆时空的美景。颐和园夜景照明必将谱写下亦科学、亦艺术、亦保护、亦发展的篇章。

1
颐和幻影

颐和园夜景照明设计框架

Design Framework for Summer Palace Landscape Lighting

自 2004 年 9 月起，天津大学承担北京科委社会发展项目"颐和园古典园林夜景照明工程技术研究与示范"，对在颐和园皇家园林实施夜景照明展开了一系列考察研究及实验工作，项目包括"古建保护""生态保护""历史美学""视觉场景"等多方向子课题。研究过程中，投入大量人力和物力进行了扎实有效的研究工作及现场实验，经过多方面有关专家的讨论和论证，形成了最终研究成果。本研究以其中"古建保护""生态保护"两个方面的成果及相关保护性建议为基础，对颐和园夜景照明设计方案中各个区域的照明方式、布灯位置、灯具选型等等进行评估分析，并参考天津大学建筑设计研究院及颐和园管理处共同编制的《颐和园文物保护规划》中相关成果及控制指标，对夜景照明设施对颐和园日间景观可能形成的影响进行评价，最终提出综合评价结果，得出兼顾古建生态保护的颐和园夜景照明设计方案。

Since September 2004, Tianjin University has undertaken the "Landscape Lighting Research and Demonstration for Classical Gardens of Summer Palace", a social development program funded by the Beijing Municipal Science & Technology Commission, carrying out a series of survey and research on and tests of the nightscape lighting for Summer Palace imperial gardens, covering "historic building protection", "ecological protection", "historical aesthetics" and "visual scenes", etc. During the process, lots of human and material resources have been used for effective research and field tests, and final research results have been made through discussions and demonstrations of relevant experts. On the basis of the achievements in "historic building protection" and "ecological protection" and relevant protection-related suggestions, in the research, assessment and analysis have been made on the lighting system, lighting layout and selection of lighting fixtures for each area in the nightscape lighting design plan for Summer Palace; by referring to relevant results and control indexes in the "Plan on Preservation of Cultural Relics of Summer Palace" jointly developed by Tianjin University Research Institute of Architectural Design & Urban Planning and Summer Palace Management Office, we have evaluated the possible impacts of lighting facilities on the daytime landscape of Summer Palace, put forward the final assessment results and made a design plan on both ecological protection and nightscape lighting of historic buildings of Summer Palace.

1
枝探万寿山

本次工程2015年得到国家文物局关于颐和园夜景照明工程的方案批复，属于对原项目的改造和升级。本成果依照的文件如下：

《北京颐和园总体规划文本》

《北京颐和园文物保护规划文本》

《古建筑木结构维护与加固技术规范》

《国家文物局关于颐和园夜景照明的批复》

《颐和园古典园林夜景照明工程技术研究与示范结题专家意见》

根据上述内容，提出本次评估的基本原则和方针：基于最小干预原则、可识别性原则、可逆性原则、与环境统一原则，贯彻以保护为前提、合理利用、加强管理的方针。

1. 照明规划思想及目标

颐和园是中国古代皇家园林的设计典范，现存园林及蕴含其中的造园哲学是中国古代珍贵的物质和文化遗产。以颐和园的造园思想、造园规划和造园手法为重要设计规划依据，充分表现中国古典园林的美学思想和艺术精髓，以及皇家园林的雍容华贵和"幽燕沉雄气"；并在继承和发扬中国古典园林美学思想的基础上，用灯光塑造出符合现代审美需求的城市夜景照明的人文景观。最大限度呈现文化精髓，保持文化尊严，实现文化体验的精品化、商业开发的科学化。

2. 照明规划范围

本规划分二期实施，分为2016年一期和2017年二期两部分。

2016年照明规划范围：增设东堤沿岸绿化照明、西堤六桥建筑节点及沿线绿化照明、南湖岛——十七孔桥——八方亭组群建筑照明，后山区域增设安全通行照明设施。具体包括霁清轩、谐趣园、后溪河、澹宁堂、苏州街、绮望轩遗址、半璧桥、南湖岛、十七孔桥、八方亭、东堤沿岸树木、西堤六桥、景明楼及西堤沿岸树木。

2017年照明规划范围：完成佛香阁、排云殿、长廊及万寿山前山建筑节点的原有灯具设施安全消隐升级改造工程，丰富园林文化观赏体验内容，在满足功能性、安全性基础上，最大限度地体现颐和园景观文化特点。具体包括万寿山及临水界面、佛香阁、画中游、景福阁周边树木、长廊、九道弯及临水建筑、文昌阁、小石桥、知春亭、石舫至北如意门沿途。

主要照明规划区域如下图所示：

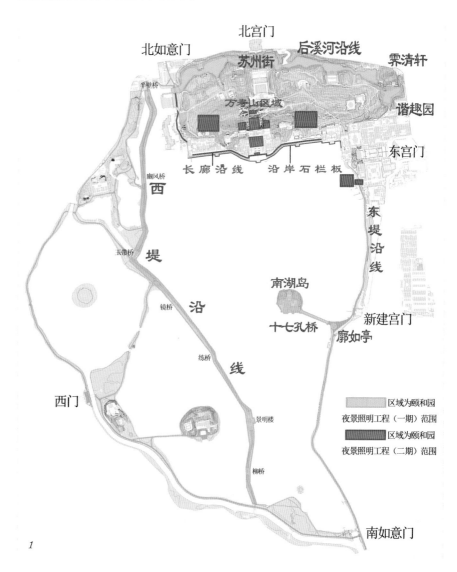

1 规划照明区域

照明设计技术要点

Technical Principles of Lighting Design

颐和园夜景照明技术除为观赏作用外，更体现对世界文化遗产的保护性评估，所以应总结的设计要点至少包括如下方面。

（1）对建筑立面绘有油饰彩画的古建筑及构筑物，根据前文相关研究成果，夜景照明中其表面照度不高于 200 lx。

（2）严格控制泛光灯的安装角度和灯具遮光角，避免对皇家园林古树群落的直接照射，古建周围散布的古树景观受到的间接照明照度不高于 5 lx，此类区域照明时间小于 2 h。

（3）园内古建筑的照明光源都需要使用低紫外线低温光源，对于古建及历史保护树木共存的区域，照明光谱严格控制在 450~650 nm 之间。

（4）保护园内动物多样性生态特征，禁止对雨燕、喜鹊等禽巢址区域直接照射，该类区域照度应严格控制在 10 lx 以下。

（5）照明设备首选安装位置应远离古建及古树文物，任何管线均须远离古建；照明条件不允许的情况下，在文物构件表面安装灯具时，禁止进行任何破损文物表面的灯具安装。

（6）公园中重要古建景观区域的照明设施，为避免其影响日间文物环境风貌，应尽量减少灯杆的数量及灯具数量。

（7）万寿山山体照明灯杆设施应做假树遮掩，使用可拆卸灯杆及升降灯杆，严格控制灯杆高度在 5.5 米以下。

（8）灯具安全等级：所使用灯具全部满足相关安全防火等级要求。

（9）选用高效节能型光源，节约能源，并降低热辐射对照明主体的影响。

The nightscape lighting technologies for Summer Palace shall be not only for ornamental purpose, but also emphasize the protection of world cultural heritages. Therefore, the technical key points shall cover, at a minimum, the following aspects:

1
佛国卧虹

(1) In accordance with the above relevant research results, the surface illumination of historic buildings and structures with decorative oil paintings on the facades in the nightscape lighting shall not be higher than 200 lx.

(2) The installation angle of floodlight and lamp shielding angle shall be strictly controlled; direct radiation to ancient trees in the imperial gardens shall be avoided; the illumination of indirect lighting for ancient trees scattered around historic buildings shall not be higher than 5 lx, and the lighting time for such areas shall be less than 2 hours.

(3) Low-UV and cold light sources shall be used for historic buildings in Summer Palace; for the areas with both historic buildings and ancient trees, the lighting spectra shall be controlled strictly within the range of 450-650 nm.

(4) Direct lighting to the areas with nests of birds including swift and magpie and beast's hideouts shall be prohibited so as to protect the ecological characteristics of animal diversity, and the illumination of such areas shall be strictly controlled under 10 lx.

(5) The installation site of lighting fixtures shall be ideally far away from historic buildings, ancient trees and cultural relics, any pipeline must be installed far away from historic buildings; if the lighting conditions do not permit, it is prohibited to install any light fixture on the surface of damaged cultural relics.

(6) The number of lamp posts and fixtures in landscape areas of important historical buildings in Summer Palace shall be minimized, so as to prevent such lamp posts and fixtures from affecting the environment and looks of the cultural relics in the daytime;

(7) The lamp posts on the Longevity Hill shall be covered by fake trees; removable and sliding lamp posts shall be used; the height of the lamp posts shall be strictly controlled under 5.5 m.

(8) Safety level of lighting fixtures: All lighting fixtures shall meet relevant requirements on fire protection.

(9) High-efficient and energy-saving light sources shall be used for energy saving and lower impact of thermal radiation on the targets.

根据上文制定的总体评价指标将颐和园保护性夜景照明设计指标归纳如下表所示，在具体照明设计中根据各区域照明主体、照明载体的不同情况，按照照明控制的各方面指标，逐项进行评价定位，最后得到既达到保护性原则，又能切实指导该区域照明设计的控制指标等级。

颐和园保护性夜景照明设计指标

照明对象	等效照量			灯具相关				光源
	照明主体表面最高照度（lx）	照射时间（h）	每天等效照量（lx·h）	照射距离	灯具安装限定（文物保护级别）	灯具外形、位置限制及日间处理措施（风貌）	安全等级	光谱限制（紫外线、红外线）
保护性动物、古树等	<10	一级、二级控制区 <2	<20	禁止对古树进行直接照明	禁止破坏生态环境的灯具安装	照明设施远离古树，并考虑日间对景观的影响	灯具及管线电器安装满足颐和园消防安全要求	450 nm
		三级控制区 <3	<30					
重点建筑及其油饰彩画	<200	<3	<500	禁止近距离照射	禁止破坏文物的灯具安装	照明设施远离古建文物，并考虑日间对景观的影响		不含有紫外线的低温光源

从总结的角度看，颐和园夜景照明工程，所采用的照明方式、光源灯具、照度控制、走线布置、安装方法、隐藏手段等，均对古建筑文物保护、园林环境保护、施工安全、日间风貌、防火防电等方面作了较为全面和充分的考虑，基本满足颐和园古典园林夜景照明评估要求。但为了更好地贯彻以保护为前提、合理利用、加强管理的方针，还要再提升、再研究、再探索。

世界遗产地保护性照明设计施工建设管理条例（建议稿）

Management Rules for Protective Lighting Design and Construction at World Heritage Sites (Proposal Manuscript)

颐和园是 1998 年 11 月在第 22 届世界遗产大会被列入《世界遗产名录》的世界文化遗产项目，截至 2017 年，中国世界遗产项目总数居世界第二，达到 52 项（自然、文化、双遗产）。"52 处世遗"彰显了一个负责任国度对人类文明的担当，与此同时，其保护与可持续发展的任务亦十分艰巨。面对国际社会让文化遗产"活"起来的强烈呼声，我们认为极有必要通过为期十几年"颐和园古典园林夜景照明与古建筑文物保护工程"的实践，归纳出对中国"52 处世遗"及数以千计的全国重点文物保护单位有指导价值的照明设计与施工的技术"导则"。

美国人文地理学家大卫·罗温索在其著作《过去即他乡》中说"过去的遗存即是历史上的他者"，也是文化上的他者。让文化遗产"活"起来，重在有展示、解读和传递遗产价值的方式。2008 年国际古迹遗址理会第 16 届大会通过的《关于文化线路的国际古迹遗址理事会宪章》指出："阐释指一切可能的、旨在提高公众意识、增进公众对文化遗产地理解的活动"。显然，这里指将文化遗产的价值更形象生动地呈现出来。联合国教科文组织编写的《世界文化遗产地管理指南》也说："每个世界遗产地都不止一个重要的故事来说明其历史：它们是如何被建造的或如何被破坏的，曾经生活在哪里的人、曾经发生过的活动和事件、遗址以前的用途和关于这些著名珍宝的传说……"在颐和园照明工程及摄影表现中借此观点，旨在表达如何改变我国遗产保护中存在的重本体修复保护、轻阐释展示的现象，因为事实上，相比于文化遗产的修复与守护，更加困难的是传承文化遗产精神之魂，使之"活在当下"。所以，颐和园的古建园林照明工程乃至拍摄的夜景照片，不仅要美，更要体现文化内涵的深度。所以，探讨并总结文化遗产地，尤其是"52处世遗"地的夜景照明设计建设技术"导则"作用将十分有效。无论从文化经济学、保护经济学出发都对世界遗产地保护与"活态"利用有如下要点，即要预测并研究世界遗产地申请成功后可能的效益与影响；要评估现有世界遗产地发展的成本和收益以检验其管理体系的效率效能；要对特定地区的世界遗产地开展专题研究，探讨文化旅游对经济和遗产保护之影响。这里不仅要从历史的视角审视文化遗产的"资本"属性，还要拓展对文化遗产的价值认知。

以下为该"导则"应遵循并坚持的总原则：坚持大遗产观与人文城市建设理念，"世遗"项目夜景照明应同时考虑古建园林、生态系统、历史与现代美学、视觉场景等科学与文化要素，此外也要结合智慧城市对智慧文博、智慧旅游的要求，逐一细化"导则"内容。本"导则"虽仅仅是一个框架，但力求列写出以颐和园夜景照明为个案的普遍性问题，避免挂一漏万。

（1）在保护原则上，真实性与完整性是建筑遗产保护的两个核心原则。《威尼斯宪章》（1964年）在强调真实性时，突出最小干预、可识别性、可逆性原则等，《奈良文件》（1994年）在很大程度上强调"真实性"应充分考虑文化的多样性；从生态安全上看，生态园林的景区"照亮"是趋势，所以要实施智慧型照明技术下的生态环境、安全防护相统一的原则。

（2）在保护方针上，坚持以保护为基础的可持续发展（含利用）理念，不断在理念和实践上升级换代，为打造世界旅游目的地，贯彻保护为前提、合理利用、强化监管的方针。

（3）在项目标准上，坚持工程前期研究为先的原则，做好同类型项目的国内外比较研究，从照明方案的分区逐项予以论证评估，乃至照明优化参数的落实，从而保障工程"全链条"的实现。达到保障历史建筑、生态园林、文博陈设及环境风貌等免受影响，实现安全防灾等指标下的照明工程建设。

1
波涛

（4）在技术法规上，建议国家文物保护主管部门牵头编制《中国世界文化遗产地照明工程建设管理办法》等法规文件，以从国家层面强化"世遗"单位照明工程的科学性且不对"世遗"地本体造成危害。

（5）在研究并应用"遗产旅游"的真实性的过程中，一方面要反对以"利用"为核心不注重遗产原真性，制造如同赝品般的光环境，另一方面要避免以片面"保护"为重，强调真实性的绝对重要，不利用遗产资源，枉顾了游人对夜间文化遗产地的向往与追求。基于遗产的视觉传达特性，本导则要求照明手段不可趋于一般化，并且不能在不经意间使照明效果失去真实性，破坏遗产的内涵价值与属性，这是相当有害的。

（6）关注照明对古建筑油饰彩画的影响，如褪色、粉化、开裂、脱落等老化形态，研究木材物质成分构成与受光照变化机理，确定适宜的光源与灯具，由于LED灯自身光谱不含紫外和红外成分，较适宜于古建筑油饰彩画照明。

（7）关注古建园林夜色照明对生态的影响，要重视光生态对生物的作用。不仅不同光源对生物的影响机理不同，同种光源对生物各生长阶段的作用也不相同，如古老树木和珍稀观赏植物夜间应杜绝一切人工照明，新栽树种周边也应避免夜间照明。

（8）关注古建及其环境景观照明工程的影响，如要做到：降低灯具设备对古建风貌的影响；电气管线布置应严格按照"所有设施远离古建及古树，对文物零破坏"的原则；保护园林动物多样性生态特征；禁止对禽兽巢址区域直接照射，该区域照度应严格控制在 10 lx 以下等。

（9）关注差异性文化旅游发展策略且鼓励公众自主参与。 不同类型的"世遗"地有不同特点，但提升服务设施的整体水平，照明尤其是夜景照明是必须研究的，无论是历史城镇型、山岳遗产地、石窟遗址型、考古遗迹型还是皇家殿宇园林型，在照明设计时，都要充分考虑其不同的脆弱性与承载力。特别要将夜景照明开放性与公众普及教育、创造高质量的城市公共文化环境相结合。

The overall principles in the "guideline": we shall follow the view of heritage and phylosophy of humanistic city construction and consider heritage protection in a holistic way, and take account of scientific and cultural elements such as ancient architecture and gardens, ecological system, historical and modern aesthetics, and visual scenes for nightscape lighting of "World Heritage". Moreover, we shall refine the content of the "guideline" on the basis of the requirements on smart relics and smart tourism. Although the "guideline" is only a framework, it shall cover universal issues with the nightscape lighting of Summer Palace as an example, so as to avoid omissions.

(1) Protection principles: authenticity and integrity are two core principles of architectural heritage protection. The Venice Charter (1964) advocates authenticity and prioritizes minimum intervention, recognizability and reversibility, etc. The Nara Document on Authenticity (1994) emphasizes to a great extent that we shall give full play to cultural diversity for "authenticity"; from the perspective of ecological safety, it is a general trend to "illuminate" the scenic areas of ecological gardens; therefore, we shall follow the principle of unification between ecological environment and safety protection by using smart lighting technologies.

(2) Protection guideline: we shall follow the phylosophy of sustainable development (containing utilization) on the basis of protection, keep upgrading the philosophy and put it into practice, ensure reasonable utilization and effective regulation under the precondition of ecological and heritage protection, so as to build Summer Palace into a world-class tourism destination.

(3) Project standards: we shall follow the principle of giving priority to preliminary project studies, conduct comparative researches on domestic and foreign projects of the same type, demonstrate and assess the lighting plan item by item, and improve the lighting parameters, thereby ensuring "full-chain" implementation of the project, avoiding affecting the historic buildings, ecological gardens, cultural exhibitions and the environment, and ensuring safety and disaster prevention.

(4) Technical regulations: it is recommended that China's authority for cultural heritage protection take a lead in preparing regulatory documents including the Management Methods for Lighting Project Construction at World Heritage Sites in China, so as to carry out lighting projects at World Heritage Sites at a national level in a more scientific way and prevent from causing any harm to the entirety of a "World Heritage Site".

(5) In the process of research on and application of authenticity of "heritage tourism", we shall, on the one hand, lay emphasis on authenticity of heritage and object to taking "utilization" as the core and creating a fake light environment, and on the other land, avoid laying stress on partial protection, emphasize the absolute importance of authenticity, and take advantage of heritage resources by considering tourists' love and pursuit of nightscape of cultural heritage sites. Based on the features of visual communication of heritage, this "guideline" requires that the lighting should mean neither to be generalized, nor to make the lighting effect lose its authenticity, nor to damage the intrinsic value and attributes, otherwise it may do considerable harm.

(6) We shall pay attention to the impact of lighting on ancient oil paintings on historic buildings, such as color fading, chalking, crazing and shedding, etc., study the composition of wood and the texture under the conditions of illumination change, and adopt suitable light sources and lighting fixtures; LED lamp is relatively suitable for illumination for ancient oil paintings on historic buildings since its spectra contain no ultraviolet and infrared.

(7) We shall pay attention to the ecological impact of nightscape lighting in gardens with historic buildings, attach importance to the effects of light ecology on living things. Different light sources have different impacts on living things and even the same light source has different impacts on different growth stages of living things. For example, no artificial lighting shall be used for ancient trees and rare ornamental plants at night; and night lighting shall be avoided in the surrounding areas of newly-planted trees.

(8) We shall pay attention to the impact of lighting projects on historic buildings and their environmental landscape. For example, we shall lower the impact of lighting fixtures on the looks of historic buildings, and strictly follow the principle of "ensuring that all lighting fixtures are installed far away from historic buildings and ancient trees and cause no damage to cultural relics" in the case of electric installation; we shall protect the ecological characteristics of animal diversity in the gardens, prohibit direct lighting to the areas with nests or hideouts, and keep the illumination of such areas strictly under 10 lx.

(9) We shall pay attention to differentiated strategies of cultural tourism development and encourage public involvement. Although different types of "World Heritage Sites" have different features, we must study how to raise the overall level of service facilities and the issue of illumination, especially nightscape lighting. We shall fully consider the vulnerability and bearing capacity of heritage sites of historic town, mountain, grotto, archaeological site, and imperial palace and garden at the time of lighting design. Particularly, we shall integrate the openness of nightscape lighting and public education, so as to create a high-quality public cultural environment.

拍摄与编后记

随着时代的发展，光线与建筑的关系也有了很大转变，从"照亮建筑"到看清它们的轮廓到细部，所获得的心理效应、给生理带来的影响，都不断得到相关的科学验证。建筑遗产较之一般公共建筑就更加复杂，单霁翔先生反复强调"文博建筑"要从"馆舍天地"走向"大千世界"，为此我们认为，在颐和园古建园林环境中不仅仅陈列、布置要用博物馆设计理念，照明环节更应遵循。为中国世界遗产地组织拍摄照片并编撰图书是中国建筑学会建筑摄影专业委员会、《中国建筑文化遗产》编辑部多次投入的事，但此次服务于世界文化遗产——颐和园的项目，实在令我们兴奋，因为它要展现的是夜景下的皇家园林，它要用摄影家之眼与文学语言再现历史。

为此，在长达数月对颐和园的拍摄中，中国建筑学会建筑摄影专业委员会的摄影师们，以敬畏之心兢兢业业工作，时值酷暑，他们克服困难，较好地完成了任务；《中国建筑文化遗产》编辑部的同人在撰文时翻阅、参考了大量

1
长河落日圆

资料，改编由颐和园管理处提供的《颐和园古典园林夜景照明古建筑文物保护及环境保护评估分析报告》时，也力求既保留技术要点、突出保护性照明的设计特色，也兼顾到如同文博陈列般照明"设计"的水准。一般博物馆室内的理想"光环境"，是自然光加人工光，或人工光仿自然光。在这样的光环境中，观众在获取有关色泽、神态等准确的信息外，更能陶冶性情。但皇家古建园林的室外照明难度大大增加，一方面从夜景照明技术上要创出一个符合自己特点并与时俱进的新天地；另一方面要展示夜景照明效果，传播好建筑影像记忆，这是当下以类多、量大、质高、快捷著称的信息时代，为公众提升绚丽观赏效果的新命题。

正基于此，在北京市颐和园管理处的指导下，承编单位《中国建筑文化遗产》编辑部的专家团队，以古建园林、皇家气派、中国文化与当代技术等理念为关键词，使拍摄工作与编辑工作上升到文化创意的层面，努力追求图书既准确记录历程，又达到耐人寻味之处，充分地用保护文化遗产的传播之力，实现文化惠民的传播理想。面对《光幻湖山——颐和园夜景灯光艺术鉴赏》一书的问世，我们要特别鸣谢以下单位。

建设单位：北京市颐和园管理处

研究单位：北京市颐和园管理处、天津大学建筑学院

设计单位：深圳市高力特实业有限公司

施工单位：北京平年照明技术有限公司

监理单位：建研凯勃建设工程咨询有限公司

同时要感谢中国文物学会会长、故宫博物院院长单霁翔为本书作序，也要感谢《颐和园古典园林夜景照明技术研究》著作及《颐和园古典园林夜景照明古建筑文物保护及环境保护评估分析报告》所提供的丰富研究成果及资料。特此致敬。

《中国建筑文化遗产》编辑部

2017年10月

Photography and Postscript

As time passes by, the relationship between light and building has undergone great changes from "illuminating buildings" to clearly displaying the contour and details of buildings, and the psychological and physiological effects of lighting to visitors have been constantly validated in a scientific way. Architectural heritages are more complicated than general public buildings. Mr. Shan Jixiang has reiterated that "historic buildings" shall go to the "boundless universe" rather than being confined to themselves. For this purpose, we insist on following the museum design phylosophy in display and arrangement in the environment of ancient architecture and gardens in Summer Palace, and particularly in lighting design. The Architectural Photography Committee of the Architectural Society of China and the China Architectural Heritage Editorial Board have taken photos and compiled books for World Heritage Sites in China for several times, but we are still very excited to provide services for the program of Summer Palace for it aims to display the nightscape of the imperial gardens and bring its history alive by means of photography and literary language.

For this end, during the past months, the photographers of the Architectural Photography Committee of the Architectural Society of China worked hard with reverence, overcame the difficulties and fulfilled their tasks in spite of the intense heat of summer; the editors of the China Architectural Heritage Editorial Board browsed and consulted lots of reference materials. When revising the "Assessment and Analysis Report on the Impacts on Nightscape Lighting for Classical Gardens of Summer Palace on Historic Buildings and Environmental Protection" provided by Summer Palace Management Office, the editors followed the principles of both retaining the key design characteristics and highlighting protective illumination, taking account of the museum-like illumination design. Generally, the ideal "indoor lighting environment" for museums is an environment with natural light and artificial light, or with simulated natural light. Such a light environment may facilitate visitors to acquire accurate information on color and expression, and more importantly, cultivate their temperament. However, outdoor lighting for imperial gardens is much more difficult, on the one hand, we need to, from the perspective of nightscape lighting, create a new horizon that is in line with the characteristics of such imperial gardens and keeps pace with the times; on the other hand, we need to give play to the effects of nightscape lighting and pass on the memories of architectural images, which puts forward a new requirement for raising the effect for the public to enjoy the nightscape in an information era featuring great variety, large amount, high quality and convenience .

Guided by Summer Palace Management Office, the expert group of the China Architectural Heritage Editorial Board, has raised the photographic and editing work to the level of cultural creativity by focusing on historic buildings, royal style, Chinese culture and modern technologies, etc., ensuring that the book not only accurately records the historical course but also provoke thoughts, with a view to protecting the cultural heritage and benefitting the people through cultural communication. We'd like to take this opportunity to express special thanks to the following units:
Summer Palace Management Office (Developer)
Summer Palace Management Office, Tianjin University School of Architecture (Researcher)
Shenzhen Gaolite Industrial Company (Designer)
Beijing Pingnian Lighting Technology Co., Ltd. (Constructor)
CABR Construction Consulting Co., Ltd. (Supervisor)

Also, we'd like to thank Shan Jixiang, President of the Chinese Society of Cultural Heritage and Director of the Palace Museum, for his writing preface for this book, and the authors of the Technical Research on Landscape Lighting in Classical Gardens of Summer Palace and "Assessment and Analysis Report on the Impacts on Nightscape Lighting for Classical Gardens of Summer Palace on Historic Buildings and Environmental Protection" for the rich research results and information contained herein.

CAH Editorial Board
October 2017